ENERGY POLICIES, POLITICS AND PRICES

AF698549

THE ENERGY-WATER NEXUS

KEY ISSUES

ENERGY POLICIES, POLITICS AND PRICES

Additional books in this series can be found on Nova's website under the Series tab.

Additional E-books in this series can be found on Nova's website under the E-book tab.

ENERGY POLICIES, POLITICS AND PRICES

THE ENERGY-WATER NEXUS

KEY ISSUES

TUCKER WEBB
AND
ELLIS PORTER
EDITORS

New York

Copyright © 2013 by Nova Science Publishers, Inc.

All rights reserved. No part of this book may be reproduced, stored in a retrieval system or transmitted in any form or by any means: electronic, electrostatic, magnetic, tape, mechanical photocopying, recording or otherwise without the written permission of the Publisher.

For permission to use material from this book please contact us:
Telephone 631-231-7269; Fax 631-231-8175
Web Site: http://www.novapublishers.com

NOTICE TO THE READER

The Publisher has taken reasonable care in the preparation of this book, but makes no expressed or implied warranty of any kind and assumes no responsibility for any errors or omissions. No liability is assumed for incidental or consequential damages in connection with or arising out of information contained in this book. The Publisher shall not be liable for any special, consequential, or exemplary damages resulting, in whole or in part, from the readers' use of, or reliance upon, this material. Any parts of this book based on government reports are so indicated and copyright is claimed for those parts to the extent applicable to compilations of such works.

Independent verification should be sought for any data, advice or recommendations contained in this book. In addition, no responsibility is assumed by the publisher for any injury and/or damage to persons or property arising from any methods, products, instructions, ideas or otherwise contained in this publication.

This publication is designed to provide accurate and authoritative information with regard to the subject matter covered herein. It is sold with the clear understanding that the Publisher is not engaged in rendering legal or any other professional services. If legal or any other expert assistance is required, the services of a competent person should be sought. FROM A DECLARATION OF PARTICIPANTS JOINTLY ADOPTED BY A COMMITTEE OF THE AMERICAN BAR ASSOCIATION AND A COMMITTEE OF PUBLISHERS.

Additional color graphics may be available in the e-book version of this book.

Library of Congress Cataloging-in-Publication Data

ISBN: 978-1-62417-753-8

Published by Nova Science Publishers, Inc. ✝ New York

Contents

PREFACE

Water and energy are inextricably linked and mutually dependent, with each affecting the others availability. Since 2009, the Government Accountability Office (GAO) has issued five reports on the interdependencies between energy and water. These reports have shown that a considerable amount of water is used to cool thermoelectric power plants, grow feedstocks and produce biofuels, and extract oil and natural gas. Some of these sources of energy may also negatively affect water quality. In addition, developing oil and gas resources can product wastewater, known as 'produced water', that must be managed or treated. Conversely, significant amounts of energy are needed to extract, transport, treat, and use water in urban areas. This book identifies key energy-water nexus issues that Congress and federal agencies need to consider when developing and implementing national policies for energy and water resources.

Chapter 1 - Water and energy are inextricably linked and mutually dependent, with each affecting the other's availability. Since 2009, GAO has issued five reports on the interdependencies between energy and water. These reports have shown that a considerable amount of water is used to cool thermoelectric power plants, grow feedstocks and produce biofuels, and extract oil and natural gas. Some of these sources of energy may also negatively affect water quality. In addition, developing oil and gas resources can produce wastewater— known as "produced water"—that must be managed or treated. Conversely, significant amounts of energy are needed to extract, transport, treat, and use water in urban areas.

GAO was asked to identify key energy-water nexus issues that Congress and federal agencies need to consider when developing and implementing national policies for energy and water resources. To conduct this work, GAO systematically reviewed its five reports to identify key nexus issues. GAO also used a content analysis of related literature and interviews with specialists to validate these themes.

Chapter 2 - The energy choices before Congress represent vastly different demands on domestic freshwater because water is used in varying amounts in most aspects of the energy sector. Transitions in the energy sector, such as the pursuit of greater energy independence and security, produce changes in how much and where the energy sector uses water. The energy sector is the fastest-growing water consumer in the United States, in part because of energy policies. Whether the federal government addresses the energy sector's rising water demand, and if so how, is one of the many energy decisions that may be considered by the 112th Congress.

Much of the growth in the energy sector's water demand is concentrated in regions with already intense competition over water. Whether the energy sector exacerbates or alleviates future water tensions is influenced by near-term energy policy and investment decisions. These decisions also may determine whether water will limit or harm U.S. capability to reliably meet the nation's energy demand. Part of the policy issue for Congress is identifying the extent of the federal role in responding to the energy sector's water use. Currently, the energy industry and states have the most responsibility for managing and meeting the energy sector's water demand.

The energy sector's water consumption is projected to rise 50% from 2005 to 2030. Projections attribute to the energy sector 85% of the growth in domestic water consumption between 2005 and 2030. The drivers of the energy sector's increasing water use are rising energy demand, greater development of domestic energy, and shifts to more water-intense energy sources and technologies. The more water used by the energy sector, the more vulnerable the energy system is to water availability. Climate change alterations of historic water patterns may exacerbate this vulnerability in some regions. While the energy sector's water demand is anticipated to rise across the United States, the West is likely to experience some of the more significant constraints in meeting this demand. Examples of regional water use concerns related to energy are shale gas production using hydraulic fracturing in many regions across the nation, some solar electricity generation in the Southwest, and current biofuel feedstock production in the High Plains.

The 112th Congress may see the issue of energy's water demand in a variety of contexts, including oversight and legislation on energy, infrastructure, environment, agriculture, public lands, climate, research, and water. Approaches for addressing energy's water demand range from maintaining the current approach to taking a range of actions. One set of available actions includes those that attempt to minimize the growth in energy's freshwater use (e.g., through promotion of water-efficient energy alternatives and energy demand management), which could be accomplished through changes to broad policies or legislation targeted at water use. Many of the possible actions to decrease water use come with energy cost, generation, and reliability penalties. Another set of actions includes measures that facilitate access to water for the energy sector. While water allocations and permits generally are a state responsibility, limited federal actions to provide water to the energy sector are possible (e.g., access to surplus water at federal reservoirs). An additional set of actions encompasses investments in data and research to inform decision making and to expand water-efficient energy technology choices. These approaches represent different potential roles and expenses for government, the energy sector, and energy consumers. Legislation in the 111th Congress proposed a variety of actions, including provisions in H.R. 469, H.R. 2454, H.R. 3598, H.R. 1145, S. 1462, S. 1733, S. 3396, and the Secure Water Act of 2009, Subtitle IV of P.L. 111-11 (H.R. 146). A significant challenge to a federal response to the energy sector's water demand is that the available options are not equally needed, attractive, or feasible across the United States.

Chapter 3 – This is the Testimony of Dr. Michael Webber, Associate Director, Center for International Energy and Environmental Policy, The University of Texas at Austin.

Chapter 4 - Energy and water are two fundamental building blocks of the U.S. economy. Power plants need water to operate. Electric water pumps move water around the country, especially in the West. Because energy and water are inextricably linked, they are in a constant interplay of supply and demand where a change in the availability one resource can impact the usage of the other and vice versa.

Especially during droughts, water scarcity can cause conflict as drinking and agriculture needs increasingly compete against energy needs.

In: The Energy-Water Nexus: Key Issues ISBN: 978-1-62417-753-8
Editors: Tucker Webb and Ellis Porter © 2013 Nova Science Publishers, Inc.

Chapter 1

ENERGY-WATER NEXUS: COORDINATED FEDERAL APPROACH NEEDED TO BETTER MANAGE ENERGY AND WATER TRADEOFFS*

United States Government Accountability Office

WHY GAO DID THIS STUDY

Water and energy are inextricably linked and mutually dependent, with each affecting the other's availability. Since 2009, GAO has issued five reports on the interdependencies between energy and water. These reports have shown that a considerable amount of water is used to cool thermoelectric power plants, grow feedstocks and produce biofuels, and extract oil and natural gas. Some of these sources of energy may also negatively affect water quality. In addition, developing oil and gas resources can produce wastewater— known as "produced water"—that must be managed or treated. Conversely, significant amounts of energy are needed to extract, transport, treat, and use water in urban areas.

GAO was asked to identify key energy-water nexus issues that Congress and federal agencies need to consider when developing and implementing

* This is an edited, reformatted and augmented version of United States Government Accountability Office, Publication No. GAO-12-880, dated September 2012.

national policies for energy and water resources. To conduct this work, GAO systematically reviewed its five reports to identify key nexus issues. GAO also used a content analysis of related literature and interviews with specialists to validate these themes.

WHAT GAO RECOMMENDS

GAO is recommending that DOE take the actions necessary to establish a program to address the energy-water nexus, with involvement from other federal agencies, as described in the Energy Policy Act of 2005. DOE agreed with the recommendation and stated that it will work with other federal agencies and experts to implement it.

WHAT GAO FOUND

As GAO's past work has shown, and other studies and specialists have confirmed, there are a number of key energy-water nexus issues that Congress and federal agencies need to consider when developing and implementing national policies for energy and water resources. Specifically:

- Location greatly influences the extent to which energy and water affect one another. For example, as GAO reported in November 2009, the impact of increased biofuel production on water resources will depend on where the feedstock is grown and whether or not irrigation is required. Consequently, it is important for Congress and federal agencies to consider the effects that national energy production and water use policies can have at the local level.
- Although technologies and approaches exist to reduce the impact of energy development on water resources and reduce the energy needed to move, use, and treat water, their widespread adoption is inhibited by barriers such as economic feasibility and regulatory challenges. In implementing energy and water policies, Congress and federal agencies will also need to be cognizant of the barriers when deciding whether to promote the adoption of these technologies and approaches.

- Making effective policy choices will continue to be challenging without more comprehensive data and research. GAO's past work has identified the need for more data and research related to the energy-water nexus, for example, to better understand hydrological processes, including aquifer recharge rates and groundwater movement. In the absence of such data and research, developing and implementing effective policies could continue to be a challenge for Congress and federal agencies.
- Improved energy and water planning will require better coordination among federal agencies and other stakeholders. GAO's work has demonstrated that energy and water planning are generally "stove-piped," with decisions about one resource made without considering impacts to the other resource. Improved planning will require federal agencies to work with one another and other stakeholders, such as state and local agencies, academia, industry, and environmental groups. Congress and some agencies have taken steps to improve coordination, but these actions are incomplete or in their early stages. For example, in the Energy Policy Act of 2005, Congress directed the Department of Energy (DOE) to establish a federal program to address the energy-water nexus, but DOE has not done so.
- Uncertainties affecting energy and water resources cannot be ignored because they could significantly affect the future supply and demand of both resources. For example, specialists GAO talked to and literature GAO reviewed identified climate change, population growth, and demographic shifts as significant uncertainties expected to exacerbate the challenges associated with managing both the supply and demand of water and energy. These uncertainties must, therefore, be accounted for when developing national policies that affect both of these resources.

ABBREVIATIONS

DOE	Department of Energy
EIA	Energy Information Administration
EPA	Environmental Protection Agency
Interior	Department of the Interior
RD&D	research, development, and demonstration

USDA U.S. Department of Agriculture
USGS U.S. Geological Survey

The Honorable Eddie Bernice Johnson
Ranking Member
Committee on Science, Space, and Technology
House of Representatives

Dear Ms. Johnson:

Water and energy are inextricably linked and mutually dependent, with each affecting the other's availability. Water is needed for energy development and generation, and energy is required to supply, use, and treat drinking water and wastewater. Water and energy are also essential to our health, quality of life, and economic growth, and consequently the demand for both of these resources continues to rise. Water is increasingly in demand to meet the needs of the public, farms, and industries, and for recreation and wildlife; and while freshwater flows abundantly in many of our nation's lakes, rivers, and streams, it is a dwindling resource in many parts of the country. Similarly, energy is increasingly in demand to support manufacturing and transportation, among other things. As the demand for water increases, the demand for energy is similarly expected to grow. While the growth rate in energy consumption in the United States has slowed over time, overall consumption continues to rise, with estimates from the Department of Energy's (DOE) Energy Information Administration (EIA) showing an expected growth of 10 percent between 2010 and 2035. To help meet this increased energy demand, domestic energy production is rising, along with its associated water usage. According to the Congressional Research Service, the energy sector has been the fastest growing water consumer in the United States in recent years and is projected to account for 85 percent of the growth in domestic water consumption between 2005 and 2030.[1] This increase in water use associated with energy development is being driven, in part, by rising energy demand, increased development of domestic energy, and shifts to more water-intense energy sources and technologies.

Since 2009, we have issued five reports on the interdependencies that exist between energy and water.[2] These reports have shown that a considerable amount of water is used to cool thermoelectric power plants,[3] grow feedstocks and convert them into biofuels, and extract oil and natural gas from geologic formations. Oil shale development would also require a great deal of water if commercial production of this energy source becomes economically feasible in the future.[4] Some of these sources of energy—for example, biofuels, which require the use of large amounts of fertilizers and pesticides to grow the feedstock—may also negatively affect water quality. In addition, development of oil and gas resources can produce large volumes of wastewater—known as "produced water"—that must be disposed of or treated to allow for its reuse. Conversely, significant amounts of energy are needed to extract, transport, treat, and use water in urban areas, additionally contributing to energy demand. Two of these reports contained recommendations, which the agencies generally agreed with and are currently in the process of implementing.

In light of the challenges we have identified in increasing the development of various domestic energy sources while ensuring they do not have detrimental impacts on water availability and quality in the future, you asked us to provide an overview of the analysis from our five energy-water nexus reports and highlight areas that Congress and federal agencies should focus on in setting and implementing water and energy policy. Based on a review of our recent work in this area as well as a review of published literature and interviews with specialists, this report examines key issues for Congress and federal agencies to consider when developing and implementing national policies for energy and water resources.

To conduct this work, we systematically reviewed our five prior reports on the energy-water nexus and identified key issues that affect the development and implementation of national policies for energy and water resources. To validate these themes, we conducted a content analysis of key literature that examines this nexus, including peer-reviewed scientific periodicals, government-sponsored research and studies, and reports from nongovernmental research organizations. We also included in the content analysis interviews with a wide range of specialists whom we identified as having expertise related to the energy-water nexus in the United States. We selected these specialists using an iterative process, soliciting additional names from each person we interviewed. From among those identified, we conducted structured interviews with specialists who could provide a broad range of perspectives on the energy-water nexus, as well as specialists whom we identified during our systematic review of studies who have analyzed (1) the energy-water nexus in general or (2) particular segments (i.e., water use by thermoelectric power plants) of the nexus. These specialists represented a variety of organizations, including federal officials; university researchers; water and energy industry representatives from groups such as the American

Water Works Association and the Electric Power Research Institute; and nongovernmental organizations, such as the Pacific Institute.[5] Representatives from federal agencies included officials, scientists, and researchers from DOE's national laboratories and EIA, the U.S. Department of Agriculture (USDA), the Department of the Interior's (Interior) U.S. Geological Survey (USGS), and the Environmental Protection Agency (EPA). We also reviewed federal laws and regulations as applicable. To conduct the content analysis of key literature and information gathered through interviews with specialists, we reviewed each study and interview and coded statements within them based on themes we developed from the GAO reports. For the purposes of reporting the results of our analysis, we used the following categories to quantify the literature and responses of specialists: "some" refers to at least two studies or specialists, "several" refers to at least five studies or specialists, and "many" refers to eight or more studies or specialists. Appendix I discusses our scope and methodology in more detail.

We conducted this performance audit from March 2012 to September 2012, in accordance with generally accepted government auditing standards. Those standards require that we plan and perform the audit to obtain sufficient, appropriate evidence to provide a reasonable basis for our findings and conclusions based on our audit objectives. We believe that the evidence obtained provides a reasonable basis for our findings and conclusions based on our audit objectives.

BACKGROUND

As our series of reports on the energy-water nexus has shown, many aspects of energy development and delivery, including resource extraction, refining and processing, generation, storage, and transportation, can affect water resources. Conversely, supplying water in an urban setting requires energy to extract, treat, and supply water to consumers. Specifically, our prior work related to the energy-water nexus focused on the following five areas:

- *Thermoelectric power plants.* In 2007, around three-fourths of the United States' electricity-generating capacity consisted of thermoelectric power plants, which rely heavily on water for cooling, as discussed in our October 2009 report.[6] Thermoelectric power plants use a fuel source—including coal; natural gas; nuclear material, such as uranium; or the sun—to boil water to produce steam, which is used to turn a turbine connected to a generator that makes electricity. As shown in figure 1, the water used to make steam (boiler water) generally circulates in a closed loop. That is, the steam from the boiler water is cooled and converted back to liquid water—referred to as condensing—in a device called a condenser and, finally, moved back to the heat source to again make steam. In typical thermoelectric plants, the water used to cool and condense the steam—known as cooling water—flows from a separate source. Power plants can use various types of water for cooling—such as freshwater or saline water—and different water sources, including surface water; groundwater; and alternative water sources, such as reclaimed water from industrial uses.

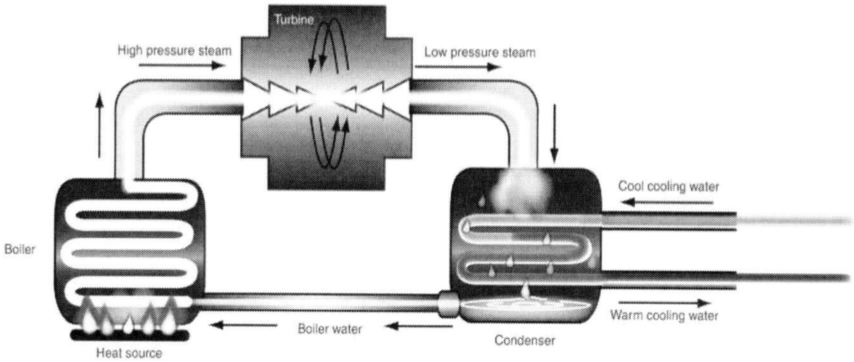

Source: GAO analysis of various national laboratory and industry sources.

Figure 1. Diagram of a Boiler Loop in a Power Plant.

- *Biofuels*. In recent years, the federal government has increasingly encouraged the use of biofuels and other alternatives to petroleum in response to concerns over the nation's dependence on imported oil, climate change, and other issues.[7] As we reported in November 2009,[8] water supply and quality can be affected by many stages of the biofuel life cycle, as can similar practices devoted to food production (see fig. 2). To cultivate feedstocks, crops can be either rainfed, with all needed water provided by natural precipitation and soil moisture, or irrigated, with at least some portion of these water requirements met through water applied from surface or groundwater sources. Water is also used in the fermentation, distillation, and cooling processes of converting the feedstock into biofuel.

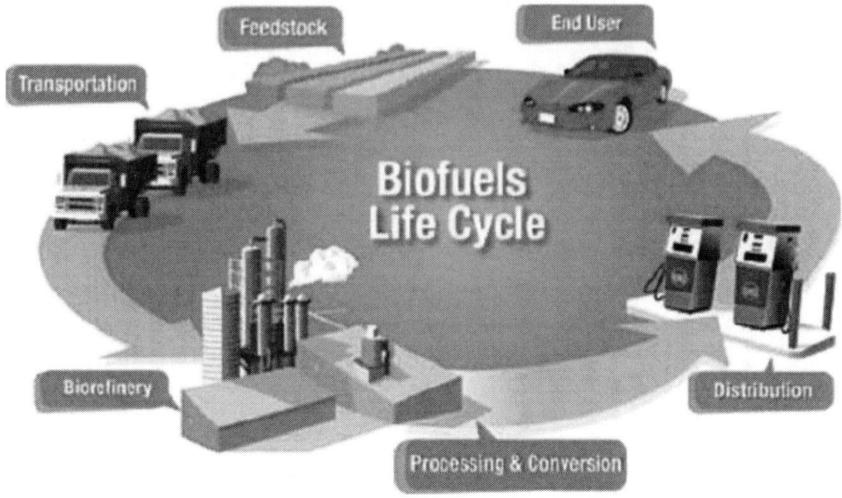

Source: DOE.

Figure 2. Biofuels Life Cycle.

- *Oil shale*. As we reported in October 2010,[9] it may become economically feasible to develop the vast amounts of oil generated from U.S. oil shale—a sedimentary rock containing solid organic material that converts into a type of crude oil when heated—and, if it does, this source could help satisfy the nation's future oil demands. Current known processes for producing oil from oil shale deposits, however, are not economically feasible—the oil costs more to produce than it could be sold for. However, if oil prices were to rise high enough to make oil shale production cost effective, and if other challenges associated with developing this resource were resolved, we reported that producing oil from oil shale could require large amounts of water. As a result, while water is expected to be available for the initial development of an oil shale industry, the ultimate potential size of any industry might be limited by water availability, among other things. Furthermore, in the absence of effective mitigation measures, water resources could be affected by ground disturbances caused by the construction of roads and production facilities; the withdrawal of water from streams and aquifers for oil shale operations; underground mining and extraction; and the discharge of waters produced from or used in oil shale operations.

- *Produced water*. The process of producing oil and gas yields several byproducts, including produced water that must be managed as part of the oil and gas operation's waste stream. Produced water may contain a variety of contaminants, including naturally occurring radionuclides, salts, and chemicals injected into some wells to help extract the oil and gas. In January 2012, [10] we reported that most produced water is reinjected into wells designed for this purpose because it requires little or no treatment and is often the least costly option,[11] but some produced water is deemed to be clean enough without extensive treatment to be used for other purposes, such as agricultural or livestock operations. However, if produced water is not appropriately managed or treated, contaminants in the water may present human health and environmental risks.

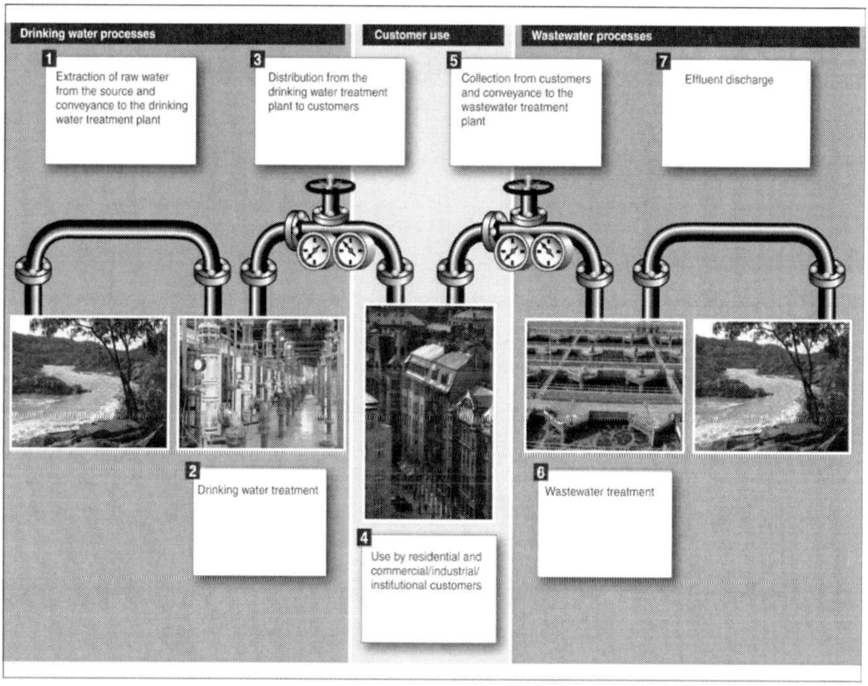

Source: GAO analysis. Photos from left to right: GAO; US EPA Photo, Eric Vance; Art Explosion; DC Water; and GAO.

Figure 3. Key Stages of the Urban Water Life Cycle.

- *Energy for water supply.* Providing drinking water and wastewater services to an urban environment involves extracting, moving, and treating water. As we described in our March 2011 report,[12] energy plays a crucial role throughout this life cycle. Energy is needed to extract raw water from the source—such as lakes, rivers, and underground aquifers—and convey it to the facility where it is treated and distributed as drinking water to customers. Furthermore, energy is needed to circulate, pressurize, and heat water for use inside households and businesses and for outdoor water-related uses by customers, such as for watering lawns. Finally, energy is required to convey wastewater to treatment facilities, treat the wastewater, and discharge the treated effluent into a receiving body of water. (See fig. 3 for the key stages in the water life cycle.)

KEY ISSUES TO CONSIDER WHEN DEVELOPING AND IMPLEMENTING NATIONAL POLICIES FOR ENERGY AND WATER RESOURCES

Our past energy-water nexus work has identified a number of key issues for Congress and federal agencies to consider when developing and implementing national policies for energy and water resources. These include (1) the varying local impacts of federal energy and water policy choices, (2) the mitigation of barriers to using innovative technologies and approaches, (3) the challenge of making effective policy choices in the absence of more comprehensive data and research, (4) the importance of coordination among governmental and nongovernmental stakeholders in order to improve planning, and (5) the attention to the uncertainties that affect energy and water resources when setting and implementing federal policies for these resources. The importance of these issues was also emphasized by the literature that we reviewed and the specialists we spoke with.

Federal Energy and Water Policy Choices Can Have Varying Local Impacts

Our past work, as well as the literature that we reviewed and the specialists we spoke with, emphasized that the amount and quality of surface and groundwater used to produce energy resources, as well as the amount and quality of water produced during oil and gas extraction, varies considerably across different locations, and this variation greatly influences how energy and water affect one another. Consequently, we believe that it will be important for Congress and federal agencies to consider the effects that national policies related to energy production and water use can potentially have at the local level. For example:

- As we reported in November 2009 on biofuels,[13] the impact of increased biofuel production on water resources will depend on where the feedstock is grown and whether or not irrigation is required. Specifically, some of the largest increases in corn acres for biofuel production are projected to occur in the Northern Plains, which relies on irrigation and is already water-constrained. Furthermore, parts of the region draw heavily from the Ogallala Aquifer, where water withdrawals for agriculture and other uses are already greater than the natural recharge rate from precipitation. As a result, developing national policies that encourage additional biofuel production could have a greater impact on local water resources in certain areas than in others.
- Similarly, because both water and energy permitting and regulations vary by state, the extent to which these two resources will affect each other depends on the state in which the development of these resources occurs. For example, as we reported in October 2009 on thermoelectric power plants,[14] some states, such as Arizona and Georgia, require thermoelectric power plant developers to obtain water use permits, while Alabama does not require such permits. As a result, the further expansion of the thermoelectric power industry in certain states can be more challenging than in others, in part, because of the need to meet state permitting and regulatory requirements for water use at those locations.
- In our January 2012 report on produced water, 15 we identified significant variations in the amount of water produced by oil and gas wells on the basis of location-related differences, such as variation in local or regional geology. For example, wells in the Barnett Shale formation in Texas, which is generally known to be a "wetter" formation than the Marcellus Shale formation in the Northeast, typically produce 3 to 4 times more water than shale gas wells in the Marcellus. We noted that even within the same formation, produced

water volumes and quality can vary greatly. The quality of produced water also varies considerably across different formations. Some produced water can be used for livestock or agricultural applications because the water is generally of high enough quality to not require extensive treatment, while other produced water may not be used in this way because it is of poor quality; for example, it may contain contaminants, such as naturally occurring radionuclides, salts, metals, oils, and production chemicals. Such water is typically re-injected into wells designed for this purpose largely because it is less costly to do so than to treat the water sufficiently to enable other uses. It will be important for federal regulators to be cognizant of these location-related variations in the quantity and quality of produced water and their related disposal and treatment implications when they develop national regulatory policies for oil and gas development.

- As some of our energy-water nexus reports have documented, the energy needed to transport water is greatly influenced by location-specific factors like topography, distance, and the type and quality of the water source. For example, in our March 2011 report on energy for water supply, 16 we reported that because pumping water accounts for 80 to 90 percent of the energy used to supply drinking water in some systems, moving water over hills and long distances can increase the level of energy consumption significantly. Consequently, when developing national policies for energy and water, policymakers need to consider the potential energy costs that can be associated with the transport and distribution of water in various settings.

In addition, the literature we reviewed and several specialists we spoke with stated that water is generally undervalued in the United States, and its prices do not reflect the true cost of providing the resource, which can result in its overuse. Some specialists also told us that the extent to which water is undervalued varies by location, with water-scarce areas being more likely to value their water resources.

Furthermore, as one specialist told us, the price customers are charged for the water they consume does not reflect all of the costs required to extract, treat, and supply the water. Therefore, consumers may be unaware of the true costs of water and more likely to waste it, which in turn leads to unnecessary energy use to produce more water.

Realizing the Benefits of Innovative Technologies and Approaches Also Depends on Mitigating Barriers to Their Use

Our past work, supported by other literature we reviewed and specialists we spoke with, has identified a variety of technologies and approaches that can reduce the impact of energy development on water resources, as well as the energy use associated with water supply, use, and treatment. However, we also identified a number of significant barriers that Congress and federal agencies will need to be cognizant of when deciding whether to promote and ensure the more widespread adoption of these technologies and approaches, as the following examples illustrate:

- As we reported in October 2009 on thermoelectric power plants,[17] advanced cooling technologies, such as dry cooling that uses air rather than water for cooling, can reduce water use at thermoelectric power plants. However, these technologies may incur “energy penalties”—that is, the energy required to power the cooling systems may reduce the plant’s net energy production to a greater extent than traditional cooling systems, potentially leading to higher electricity prices. In addition, advanced cooling technologies can have capital costs that are up to 4 times as expensive as traditional cooling systems, and they may operate less efficiently in dry, arid locations, among other concerns. As a result, policymakers seeking to encourage the adoption of these technologies by thermoelectric power plants with the goal of reducing water use will also have to consider ways to increase their economic feasibility and reduce their costs to plant operators and electricity consumers.
- As we reported in our 2009 reports on thermoelectric power plants and on biofuels,[18] the use of alternative water sources, such as treated effluent or groundwater unsuitable for drinking or irrigation, can reduce dependence on freshwater at thermoelectric power plants and biofuel refineries but may present drawbacks, such as adverse effects on cooling equipment, the need for expensive treatment equipment, and regulatory challenges. For example, power plants must comply with a number of water quality and air regulations, and the presence of certain pollutants in alternative water sources can make compliance more challenging. Furthermore, the physical layout of power plants and biorefineries may need to be changed to accommodate the use of alternative water sources. Consequently, if the goal of national water

and energy policy is to encourage the use of alternative water sources instead of freshwater sources for the production of energy, then it would also be important to focus on developing better information about the tradeoffs of using alternative water sources.

- As we discussed in our 2009 report on biofuels, [19] agricultural conservation practices can reduce the potential effects of increased biofuel feedstock cultivation on water resources, but there are barriers to their widespread adoption. For example, conservation tillage practices—such as "no-till" systems or reduced tillage systems, where the previous year's crop residues are left on the fields and new crops are planted directly into these residues—can help reduce soil erosion. Research conducted by USDA has shown a substantial reduction in cropland erosion since 1985, when incentives were put in place to encourage the adoption of conservation tillage practices. However, many farmers do not have the expertise or training to implement certain agricultural practices, and some practices may be less suited for some places. For example, farmers usually need a year or more of experience with reduced tillage before they can achieve the same crop yields they had with conventional tillage, and the amount of agricultural residue that can be removed varies by region and even by farm. Consequently, a national policy encouraging additional biofuel production would benefit from continued education and outreach provided by the federal government to help farmers better understand the advantages of adopting such conservation practices.
- More energy-efficient equipment could reduce the energy required to move and treat water, but the adoption of such technology may be hindered by the costs of retrofitting water treatment facilities and other obstacles, as we discussed in our March 2011 report on energy for water supply.[20] For example, the use of variable frequency drives at water treatment facilities, which allow operators to accommodate variations in water flows and run pumps at lower speeds, can reduce energy use by 5 to 50 percent or more. However, installing the drives can be cost prohibitive, and they are not necessarily well suited in all instances, such as when water flow is relatively constant. Therefore, federal policies to reduce energy consumption would also have to recognize the challenges faced by treatment facility operators in reducing energy consumption at their facilities and explore opportunities to assist them with this transition.

- As we also reported in March 2011 on energy for water supply, [21] water conservation is an approach to reducing the energy needed for the urban water lifecycle. Specifically, decreased customer water use could directly translate into energy savings, and water conservation also reduces the amount of energy used to convey, treat, and distribute water to customers. Similarly, energy conservation measures would also decrease water use. Therefore, when considering the development of additional energy and water conservation measures, policymakers may want to consider not only those that can be used by utilities but also those that can be used by consumers.

Making Effective Policy Choices Will Continue to Be a Challenge in the Absence of Comprehensive Data and Research

Our past work, supported by other literature we reviewed and specialists we spoke with, identified a dearth of key data and research related to energy and water. We believe that developing and implementing effective policies will continue to be a challenge for Congress and federal agencies in the absence of such data and research. Support for this data development will be essential to improved decisionmaking, as the following examples demonstrate:

- As a number of our energy-water nexus reports have found, data on the availability of freshwater and alternative water supplies are limited, and the lack of such data makes it challenging to fully assess the impact that particular energy policy choices will have on water resources. Several specialists we spoke with raised concerns that such data will be even harder to obtain in light of decreased funding for USGS's monitoring efforts. Specifically, as we discussed in our October 2009 report on thermoelectric power plants,[22] some state regulators and water experts expressed concern about a reduction in the number of streamflow gauges, which they said may make evaluating trends in water availability and water planning more difficult in the future. Without accurate data on water availability, water planning and decisions involving allocation of water resources— including power plant permitting decisions—may be less informed, according to regulators and experts we spoke with. In addition, our past work and some specialists we spoke with identified a need for more data on the quantity and quality of existing water supplies. For example, in our October 2010 report on oil shale,[23] we reported that data on the water quantity and quality baseline conditions of groundwater and surface water supplies are not sufficient for monitoring the potential impacts of oil shale development in the future. We believe that effective decisions about withdrawing water from existing supplies for energy production cannot be made without first understanding how much water is actually available and the quality of these supplies, and that the federal government plays a key role in generating these kinds of important data.

- Additional research and data on hydrological processes are needed. Our reports and discussions with specialists identified a number of areas needing more research, including the interactions between groundwater and surface water, to understand the possible transport of contaminants from energy development, aquifer recharge rates and groundwater movement, and precipitation runoff, as well as how much water is available for and used by crops. [24] For example, as we previously discussed in our November 2009 report on biofuels,[25] several specialists and agency officials have stated that research into hydrogeological processes is needed to understand the rate at which aquifers are replenished and the impacts of increased biofuel production on those aquifers. Although research suggests there should be sufficient water resources to meet future feedstock production demands at a national level, increases in this production may lead to significant water shortages in certain regions. Even in typically water-rich states, such as Iowa, concerns have arisen over the effects of increased biofuel production, and research is needed to assess the hydrology and quality of a state's aquifers to help ensure the state is on a path to sustainable biofuel production. Consequently, not only is there a need for baseline data on existing water supplies, but there is also a need for more information on how quickly such supplies are replenished and how they are affected by contaminants, so that as Congress and federal agencies are developing and implementing national energy policies they can ensure that they are considering sustainable water supplies for energy production over the long term.
- Research to determine how new technologies will affect the energy-water nexus has not been conducted to demonstrate the effects of these technologies at commercial scales. For example, according to many specialists we spoke with and some studies we reviewed,

implementing carbon capture and sequestration technologies would consume large amounts of freshwater and could affect the quality of nearby water supplies,[26] but the extent of such effects is not well known.[27] In the case of biofuels, next generation feedstocks for biofuels—such as algae or cellulosic materials, including stalks, stems, branches, and leaves—have the potential for fewer negative effects on water resources, but the magnitude of these effects remains largely unknown because these feedstocks have not yet been grown on a commercial scale. For example, as we discussed in our November 2009 report on biofuels,[28] conversion of cellulosic feedstocks is expected to use less water compared with conventional feedstocks in the long run. However, commercial-scale production has not yet been demonstrated; therefore, any estimates on water use by cellulosic biorefineries are simply projections at this time. Focusing only on certain potential benefits of new technologies without understanding the full impacts of such technologies can have unintended consequences, a concern voiced by several of the specialists we spoke with. Therefore, along with federal support for data and research into new technologies it will also be essential for Congress and federal agencies to support data and research to gain a better understanding of how adopting these technologies will affect both energy and water resources as well as the environment more broadly.

- Data on water or energy use by existing technologies and research into how to optimize these technologies is needed, according to our past reports and specialists we spoke with. For example, as we previously discussed in our November 2009 report on biofuels,[29] research into biorefinery cooling systems could help reduce the water used during conversion of feedstocks into biofuels. According to DOE officials, research and development is also needed to improve hybrid wet-dry cooling systems for thermoelectric power plants. In addition, several specialists told us that analyses are needed on the amount of energy or water used throughout the entire life cycle of supplying water and producing energy. For example, in the case of thermoelectric power, a life-cycle analysis would look beyond just the water consumed in cooling towers and would assess the amount of water needed to mine, refine, transport, and burn the coal; however, such an analysis would require a level of detail beyond that which is currently available for water data, according to specialists we spoke with. Additional data gathering and research on existing technologies can help energy and water suppliers better understand the true impacts of existing technologies on energy and water resources and consider those impacts in comparison with the potential advantages of new technologies. Data gathering and research is another area where we believe that federal agencies could play an important role and assist with such analyses.

In addition, research into and development of better models and decision support tools would help inform decisions related to the energy-water nexus, according to some reports we reviewed and specialists we spoke with. For example, as one specialist told us, while there are many national climate change models, regional models are needed to increase the accuracy of projections. Efforts are underway to develop downscaled models, but data may not be available to populate the models.[30] In addition, EPA officials told us that the U.S. Global Climate Research Program, which coordinates and integrates federal research on global environmental changes and their implications for society, is considering regional modeling and studies for cross-agency climate change in the future; however, these plans are not yet final. In the absence of improved models and decision support tools, Congress and federal agencies may be making decisions that affect energy and water supplies without fully understanding the impact of these decisions.

Coordination among Governmental and Nongovernmental Entities Is Key to Improved Planning

Our past work, the literature we reviewed, and specialists we spoke with noted that, in general, energy and water planning are "stove-piped" and frequently split across federal, state, and local levels, which results in decision making that does not adequately account for the interactions between energy and water. Improved energy and water planning will require federal agencies with oversight of these resources to coordinate with one another as well as with other stakeholders, such as state and local agencies, academia, industry, and environmental groups, as the following examples demonstrate:

- Our previous reports have highlighted the need for a better understanding of the energy-water nexus, and two of these reports recommended that federal agencies collaborate on research and data collection efforts related to it. Specifically, in our October 2009 thermoelectric power plant report,[31] we recommended that EIA and USGS establish a process for regularly coordinating with each other, other federal agencies, and water and electricity industry experts, among others, to improve the overall quality of data collected on water use from power plants. DOE and Interior generally agreed with our recommendation, and both agencies are currently in the process of implementing it. Specifically, EIA and USGS have held two meetings to discuss such matters. USGS organized the initial meeting in November 2010, and EIA hosted the second meeting in March 2012. Invitees to the meetings also included representatives from federal agencies and national laboratories. USGS is scheduled to host the next meeting in October 2012. In addition, in our October 2010 oil shale report,[32] we recommended that the Secretary of the Interior coordinate with DOE and state agencies with regulatory authority over water resources on oil shale data collection and modeling efforts, and provide a mechanism for water-related research collaboration and results sharing. Interior generally agreed and has begun to take some steps to implement the recommendation. Specifically, Interior stated that it was working to improve coordination with DOE and state agencies with regulatory authority over water resources. The literature we reviewed and many specialists we spoke with concurred that federal agencies should collaborate to better understand the energy-water nexus. As some specialists told us, without such coordination and collaboration between federal agencies, it will be difficult to ensure that energy-water tradeoffs are properly considered.

- Our previous reports recognized that adequately addressing the energy-water nexus will require federal agencies to seek input from relevant stakeholders outside of the federal government. Specifically, we recommended in both our thermoelectric power plant and oil shale reports that federal agencies collaborate with entities outside of the federal government. In response to the recommendation in our October 2009 report on thermoelectric power plants, [33] DOE and Interior generally agreed with our recommendation and, as described earlier, EIA and USGS have since held two meetings. In addition to the federal agencies that participated, nonfederal invitees to the meetings included industry groups, such as the Nuclear Energy Institute; universities, such as Southern Illinois University; and environmental groups, such as The Nature Conservancy. In our

October 2010 oil shale report,[34] Interior generally agreed with our recommendation to collaborate with entities outside of the federal government and has begun to take some steps to implement the recommendation. Specifically, Interior stated that it was working with state agencies that have regulatory authority over water resources. The literature we reviewed and many specialists we spoke with also believed collaboration with stakeholders outside the federal government is needed. For example, in its 2006 report to Congress,[35] DOE reported that the lack of integrated energy and water planning and management has already affected energy production in many basins and regions across the country, and that collaboration on energy and water resource planning is needed among federal, regional, and state agencies, as well as with industry and other stakeholders. As some specialists told us, state and local agencies are primarily responsible for allocating and managing water resources; therefore, energy-water nexus discussions need to involve officials from other levels of government. Similarly, public utility commissions make energy decisions in many states, and the role of these commissions is important to consider when addressing the energy-water nexus.[36] Furthermore, as DOE officials noted, many political boundaries between states are formed by water bodies. Therefore, multiple counties and states can be involved in water management decisions, which can lead to conflicts and be an impediment to water planning. Because of the various stakeholders with a role in managing energy and water, we believe that it will be important for federal agencies to improve coordination including with their nonfederal partners in order to enhance energy and water planning and put the United States on a path that is more sustainable for the long run.

- Many specialists told us that, because of the stove-piped nature of energy and water issues in the federal government, a top-down emphasis on collaboration among federal agencies and with other groups outside of the federal government is needed. Some specialists pointed to a recent memorandum of agreement between three federal agencies on unconventional oil and gas research as a good model. Specifically, in March 2011, the White House released the *Blueprint for a Secure Energy Future*, a comprehensive plan to reduce America's oil dependence, save consumers money, and make the United States the leader in clean energy industries. The Blueprint instructed the federal government to conduct research to examine the impacts of hydraulic fracturing on water resources. To fulfill this directive, DOE, Interior, and EPA signed a memorandum in April 2012 describing their plans for multiagency collaboration on unconventional oil and gas research. According to the specialists we spoke with, direction from the White House is what led to action on the part of the agencies and the signing of the memorandum, and they believe a similar high-level push would be needed to spur meaningful collaborative action to address other energy-water nexus issues.

Recognizing the need for better coordination, Congress and some federal agencies have taken or are beginning to take some steps on energy-water nexus issues, but many of these actions are either in their early stages or incomplete, as the following examples illustrate:

- With the Energy Policy Act of 2005, Congress directed the Secretary of Energy, through DOE's Office of Science,[37] to carry out a number of programs related to a wide variety of energy sciences. In particular, the act directs the Secretary to carry out a program to address the energy-water nexus and assess the effectiveness of existing programs at DOE and other federal agencies to address the nexus.[38] The provision also directs the Secretary to consult with the Administrator of EPA, the Secretary of the Interior, the Chief Engineer of the Army Corps of Engineers, the Secretary of Commerce, the Secretary of Defense, and other federal agencies as appropriate, to carry out the provision. DOE officials told us, however, that to date they have neither received nor requested any funding specifically designated to carry out the provision, nor has the agency requested funds for this purpose for fiscal year 2013. Furthermore, according to DOE officials, the Office of Science is not the appropriate office within DOE to carry out the program because the office conducts research on the fundamentals of science. Instead, officials stated that the Office of the Under Secretary, which oversees work in areas such as energy efficiency and renewable and fossil energy, would be the more appropriate office to conduct such research.[39] Regardless of which office is best suited for this activity, by not carrying out this provision, we believe that DOE is missing an important opportunity to provide information that could possibly help fill some of the research and data gaps we and others have identified.

- Congress is also considering pending legislation related to energy and water that calls for enhanced collaborative efforts. A House bill, the Energy and Water Research Integration Act of 2012, would direct the Secretary of Energy to (1) integrate water considerations into energy research, development, and demonstration (RD&D) programs and projects; (2) develop a strategic plan identifying RD&D needs for those programs and projects; (3) collaborate with other federal agencies, within DOE, and with nongovernmental entities in developing the plan; and (4) develop an Energy-Water Architecture Council consisting of federal agencies and nongovernmental entities to promote data collection, reporting, and technology innovation. 40 Another House bill, the Coordinating Water Research for a Clean Water Future Act of 2012, would, among other things, direct the President to begin implementation of a National Water and Research Development Initiative to, among other things, conduct research on how to ensure the systematic and coordinated collection of publicly available data on regional and national water resources and provide for interagency coordination of federal water research and development.[41]
- Some federal agencies have recognized the importance of collaboration on the energy-water nexus and have acted accordingly. For example, in April 2012, EPA proposed six principles to foster collaboration in the water and energy sectors to meet water and energy needs nationally and locally. One of the principles called for key stakeholders—governments, utilities, manufacturers, and consumers—in both sectors to move toward integrated energy and water management from source, production, and generation, to the end user. Actions to support the principle include breaking down institutional barriers, improving transparency, and maximizing efficiencies; encouraging government agencies to look across missions; and developing partnerships between government and service providers.
- DOE, led by Sandia National Laboratories, held a series of workshops on energy and water issues in 2005 and 2006, according to DOE's 2006 report to Congress. The effort, known as the Energy-Water Roadmap process, included representatives from a broad range of groups, including environmental organizations; policy and regulatory groups; industry associations; and federal, state, and tribal government agencies. It was intended to help DOE and the United

States assess current energy and water issues and concerns and identify appropriate interactions and coordination approaches. As part of this effort, participants at the workshops identified (1) gaps between current federal and state energy and water research and management programs and (2) major science and technology research and development steps necessary to address the challenges and gaps. According to the website for the Sandia National Laboratories, the Energy-Water Roadmap process was expected to result in a report summarizing needs; prioritization criteria; major gaps; innovative technical approaches and associated research needs; research and development priorities and strategies; and associated policy, regulatory, and economic assessments. However, as of September 2012, no report has been issued.

Uncertainties that Affect Energy and Water Must Be Considered in Setting Federal Policies Related to These Resources

Uncertainties—including the future makeup of the nation's energy portfolio and the potential impacts of climate change—must be considered when developing and implementing national energy and water policies, according to our past work on the energy-water nexus, studies we reviewed, and specialists we spoke with. As the following examples demonstrate, these uncertainties could significantly affect the future supply and demand of both energy and water, and therefore are external factors that must be accounted for when developing national policies:

- The magnitude of the impacts on water resources stemming from the nation's future energy use will vary depending on what sources of energy are pursued. For example, developing unconventional energy sources, such as shale oil and shale gas,[42] could have significant impacts on the quality and quantity of water resources, but the magnitude of these impacts is unknown because of uncertainty about the future scale and scope of shale oil and gas development. Similarly, if oil shale becomes economically feasible to produce on a commercial scale, the industry could have significant impacts on water quality and quantity, but the magnitude of these impacts is unknown because of technological uncertainties and because the size of a future oil shale industry is unknown, as we discussed in our October 2010 report on oil shale.[43] In addition, a recent report from DOE's National Renewable Energy Laboratory looked at a variety of scenarios for the increased use of renewable energy sources to generate electricity. For example, the laboratory reported that if the country were able to switch to 80 percent renewable electricity by the year 2050, such a switch could reduce the power sector's annual water use by approximately 50 percent.[44] However, renewable energy sources also vary in their water use, so the type of renewable energy we use in the future will have varying impacts on water supplies. For example, solar photovoltaic panels and wind turbines consume minimal water during normal operation.[45] However, concentrating solar power plants that use wet cooling could significantly increase water demand, consuming up to twice as much water per unit of electricity produced as traditional fossil fuel power plants.[46] Concerns with concentrating solar power plants are particularly acute in the Southwest—a prime location for siting these facilities because of abundant sunshine—because water supplies in the region are already limited.[47] Consequently, not fully understanding and accounting for the potential differences in the future energy portfolio's impact on water supply and quality when developing a national energy policy may result in negative unintended consequences.

- As we previously discussed in our March 2011 report on energy for water supply,[48] to address growing concerns about emerging contaminants and nutrients in the nation's water bodies, additional or more stringent regulatory standards could increase the energy demands of treatment processes in the future, but the degree to which energy use would increase depends on the regulations that are implemented. Specifically, promulgation of more stringent standards would most likely require additional levels of treatment, and the use of more energy-intensive technologies, such as ozonation and membrane filtration,[49] may be necessary to meet such new standards.[50] Regulatory changes could also increase energy demands at other stages of the urban water life cycle. For example, higher standards for effluent discharge from wastewater treatment plants could increase the energy required for treatment of the effluent. Furthermore, stricter water quality standards for receiving waters could necessitate that more plants employ advanced treatment standards, resulting in increased energy use for the additional treatment or to pump effluent farther away to other waters. As a result, policymakers will need to consider the energy inputs that will be required to meet new water quality standards when developing these regulatory choices.

According to the literature we reviewed and specialists we spoke with, climate change, population growth, increased competition for resources, and demographic shifts are expected to exacerbate the challenges associated with water and energy supply and demand, and shifts in any of these areas are expected to increase demand for both of these resources. Moreover, the effects of climate change are expected to vary by location and, in some locations, are expected to increase demand for both energy and water resources while simultaneously decreasing water supplies. According to the literature we reviewed, higher temperatures from climate change are expected to lead to additional demand for air conditioning and, therefore, electricity. This increased electricity demand will, in turn, lead to increases in water consumption associated with power generation. However, at the same time, climate change is expected to change the quantity and reliability of water supplies so that less water may be available in some regions, thereby resulting in reduced water supplies for use by the energy sector, according to some specialists we spoke with. In addition, as one specialist told us, higher temperatures from climate change will produce more evaporation from water reservoirs and other bodies of water, such as the Great Lakes, which can produce significant water losses. Furthermore, the literature we reviewed and specialists we spoke with noted that the impacts of climate change are uncertain with some areas receiving more precipitation than they currently experience, rather than drought.

Problems associated with climate change are only exacerbated by population growth and competition for water resources. Specifically, more people will consume more water, increasing the municipal sector's water demand. To meet these increasing demands, some states, especially those in areas that are already water stressed, such as Texas, have pursued alternative sources of water, such as desalinated water, which are more energy-intensive than traditional groundwater and surface water supplies. In addition, because of a warmer climate and decreased precipitation, farmers are expected to withdraw more water to irrigate crops. Minimum water levels are also necessary for other uses, such as recreation and industry, as well as to support wildlife and maintain ecosystems. Furthermore, demographic shifts, such as migration to the hot, arid Southwest, could place additional demands on both energy and water supplies. In light of all of these uncertainties, some specialists told us that water and energy policies must be resilient and flexible enough to adapt to changing circumstances. For example, these specialists told us that because of the uncertain effects of climate change, the development of plans and policies should be adaptable to account for variability over time and by location.

CONCLUSION

The growth in water and energy demands for development and other uses is occurring at a time when the nation's supplies are stressed by a growing population, a variety of new and changing uses, and environmental challenges such as climate change.

A number of agencies have responsibility for managing specific aspects of the energy-water nexus, including DOE, EPA, USDA, and Interior, but these agencies do not consistently or strategically collaborate on these inextricably linked issues to ensure a harmonized approach to energy and water resource planning. This lack of coordination often results in a stove-piped approach to managing these resources that does not take into account all the possible tradeoffs and interrelationships.

Some agencies have taken steps to examine this issue within their own agency, such as DOE and Interior's jointly organized meetings to improve thermoelectric power plant data in response to a recommendation from our October 2009 report. However, in general, federal efforts remain uncoordinated. Moreover, there is a need for enhanced research and data efforts that could benefit from greater interagency cooperation. The Energy Policy Act of 2005 requires DOE to implement a program of research, development, demonstration, and commercial action to address energy and water issues and assess existing federal programs, but DOE has not yet implemented this program.

In not carrying out this directive, DOE is missing an opportunity to provide information that could help Congress, other federal agencies, and the public better understand the key energy-water nexus issues identified in our series of reports and take coordinated action to protect these invaluable resources.

Recommendation for Executive Action

To help address some of the research and data gaps that we and others have identified related to the tradeoffs associated with the energy and water nexus, and to ensure collaboration to address the nexus, we recommend that the Secretary of Energy take the actions necessary to establish a program to address the energy-water nexus, with involvement from other federal agencies as described in the Energy Policy Act of 2005.

Agency Comments and Our Evaluation

We provided a draft of this report to USDA, DOE, Interior, and EPA for review and comment. In its written comments, reproduced in appendix II, DOE agreed with our recommendation.

Specifically, DOE stated that it already has a number of research activities underway related to the energy-water nexus, such as energy-water data collection and modeling. In addition, DOE stated that it will initiate more assertive engagement with program managers through an internal workshop to discuss existing activities and clarify priority areas for further data collection and analysis. As part of its efforts, DOE stated that it will continue to engage other agencies and collaborate with experts.

EPA provided technical comments, which we incorporated as appropriate. In its technical comments, EPA stated that the agency supported the findings of the report, but noted that more emphasis should be placed on the changes in population demographics that stress the supply of energy and water. Specifically, the agency identified increasing population shifts to coastal areas, desalinization, increased use of air conditioning, and agricultural irrigation as stressors of particular concern.

We agree that these areas could affect the energy-water nexus, as we discuss in our report. EPA also identified the need for a comprehensive systems analysis of energy, water, and agricultural cycles and their feedback loops to help avoid unintended consequences. USDA and Interior did not provide any comments.

As agreed with your office, unless you publicly announce the contents of this report earlier, we plan no further distribution until 30 days from the report date. At that time, we will send copies to the Secretary of Energy, the appropriate congressional committees, and other interested parties. In addition, the report will be available at no charge on the GAO website at http://www.gao.gov.

If you or your staff have any questions about this report, please contact Anu K. Mittal at (202) 512-6100 or mittala@gao.gov or Frank Rusco at (202) 512-3841 or ruscof@gao.gov. Contact points for our Offices of Congressional Relations and Public Affairs may be found on the last page of this report. GAO staff who made key contributions to this report are listed in appendix III.

Sincerely yours,

Anu K. Mittal
Director,
Natural Resources and Environment

Frank Rusco
Director,
Natural Resources and Environment

APPENDIX I: OBJECTIVES, SCOPE, AND METHODOLOGY

To conduct this work, we systematically reviewed GAO's five prior reports on the energy-water nexus to develop the key issues that Congress and federal agencies need to consider when developing and implementing national policies for energy and water resources.

To validate these themes, we conducted a content analysis of 11 key studies and reports that examine this nexus, including peer-reviewed scientific periodicals, government-sponsored research and studies, and reports from nongovernmental research organizations.

We also included in the content analysis 19 interviews with a wide range of 37 specialists whom we identified as having expertise related to the energy-water nexus in the United States. For the purposes of our interview analysis, each interview represents the views of one specialist even if more than one specialist was present at the interview. We selected these specialists using an iterative process, soliciting additional names from each person we interviewed.

From among those identified, we conducted structured interviews with specialists who could provide us with a broad range of perspectives on the energy-water nexus as well as specialists whom we identified during our systematic review of studies who have analyzed (1) the energy-water nexus in general or (2) particular segments (i.e., water use by thermoelectric power plants) of the nexus.

These specialists represented a variety of organizations, including federal officials; university researchers; water and energy industry representatives from groups such as the American Water Works Association and the Electric Power Research Institute; and nongovernmental organizations, such as the Pacific Institute.[1]

Representatives from federal agencies included officials, scientists, and researchers from the Department of Energy's national laboratories and Energy Information Administration; the Environmental Protection Agency; the Department of the Interior's U.S. Geological Survey; and the U.S. Department of Agriculture. We also reviewed federal laws and regulations as applicable.

We conducted a content analysis on the key literature and interviews with specialists using NVivo software. Content analysis is a methodology for structuring and analyzing written material.

Specifically, using our systematic review of GAO's energy-water nexus reports to identify key issues that Congress and federal agencies need to consider when developing and implementing national policies for energy and water resources, we reviewed each study and interview and coded statements within them based on themes we developed from the GAO reports.

The coding was conducted independently by two GAO analysts after checking for intercoder reliability.

We developed agreement statistics and discussed and resolved any discrepancies in coding. We used the following categories to quantify the literature and responses of specialists: "some" refers to at least two studies or specialists, "several" refers to at least five studies or specialists, and "many" refers to eight or more studies or specialists.

We conducted this performance audit from March 2012 to September 2012, in accordance with generally accepted government auditing standards. Those standards require that we plan and perform the audit to obtain sufficient, appropriate evidence to provide a reasonable basis for our findings and conclusions based on our audit objectives.

We believe that the evidence obtained provides a reasonable basis for our findings and conclusions based on our audit objectives.

APPENDIX II: COMMENTS FROM THE DEPARTMENT OF ENERGY

Department of Energy
Washington, DC 20585

August 28, 2012

Mr. Frank Rusco
Director
Natural Resources and Environment
U.S. Government Accountability Office
441 G Street, N.W.
Washington, D.C. 20548

Dear Mr. Rusco:

The Department of Energy (DOE) welcomes the opportunity to respond to the recommendation by the U.S. Government Accountability Office (GAO) for the agency to act to address the energy-water nexus. DOE recognizes America's national interest in sound management of all natural resources, including water and energy. The issues raised by the nexus between energy and water span beyond any one technology program at the DOE, and, in fact, they reach virtually every energy program.

Current DOE research activities relevant to the energy-water nexus include but are not limited to lifecycle cost analysis, energy-water data collection and modeling (including for transmission planning purposes), development of more water-efficient energy technologies and more energy-efficient technologies involving water, integration of renewable energy in the electric power grid, drought and climate change vulnerability assessments within the energy sector. Because of the variety of roles that water plays across different energy technologies, DOE can best address the issues by incorporating consideration of water directly into the specific energy programs, rather than creating a separate silo focused on the energy-water nexus.

Recognizing that there is much more yet to analyze, DOE accepts the GAO recommendation that additional effort be committed to help address some of the gaps in data and analysis for decision-makers. Because multiple DOE research program offices have an interest in the water input and output attributes of various energy technologies, these experts are already in dialogue, and DOE will benefit from improved awareness and communication about energy-water interdependencies. With the GAO recommendation, DOE will initiate more assertive engagement with program managers through an internal workshop to surface any further existing activities and clarify priority areas for further data collection and analysis. In addition to integrating considerations about the energy-water nexus into activities across the applied energy research and development programs, DOE will also continue to engage other agencies relevant to the energy-water nexus in order to coordinate and collaborate with experts who

Printed with soy ink on recycled paper

have complementary interests and information resources and will expand these efforts where appropriate.

DOE appreciates the contribution GAO has made to an important policy dialogue through its series of reports on energy and water, and particularly in this capstone report highlighting some of the persistent challenges and opportunities for the Federal government.

Sincerely,

David B. Sandalow
Under Secretary of Energy (Acting) and
Assistant Secretary for Policy and
International Affairs

End Notes

[1] Water consumption refers to the portion of the water withdrawn that is no longer available to be returned to a water source, such as when it has evaporated. Energy production (which includes biofuel production), together with thermoelectric power, are the second largest consumer of water in the United States, accounting for approximately 11 percent of water consumption in 2005. Irrigation was the largest consumer, at approximately 74 percent. (Elcock, D., "Future U.S. Water Consumption: The Role of Energy Production," *Journal of the American Water Resources Association*, vol. 46, no. 3 (2010): 447-460.). However, according to the U.S. Geological Survey, in terms of water withdrawal, thermoelectric power was the largest source of water withdrawals (49 percent) in 2005, followed by irrigation at 31 percent. Water withdrawal refers to water removed from the ground or diverted from a surface water source, such as an ocean, river, or lake.

[2] GAO, *Energy-Water Nexus: Improvements to Federal Water Use Data Would Increase Understanding of Trends in Power Plant Water Use*, GAO-10-23 (Washington, D.C.: Oct. 16, 2009); GAO, *Energy-Water Nexus: Many Uncertainties Remain about National and Regional Effects of Increased Biofuel Production on Water Resources*, GAO-10-116 (Washington, D.C.: Nov. 30, 2009); GAO, *Energy-Water Nexus: Amount of Energy Needed to Supply, Use, and Treat Water Is Location-Specific and Can Be Reduced by Certain Technologies and Approaches*, GAO-11-225 (Washington, D.C.: Mar. 23, 2011); GAO, *Energy-Water Nexus: A Better and Coordinated Understanding of Water Resources Could Help Mitigate the Impacts of Potential Oil Shale Development*, GAO-11-35 (Washington, D.C.: Oct. 29, 2010); and GAO, *Energy-Water Nexus: Information on the Quantity, Quality, and Management of Water Produced during Oil and Gas Production*, GAO-12-156 (Washington, D.C.: Jan. 9, 2012).

[3] Thermoelectric power plants use a fuel source—for example, coal, natural gas, nuclear material such as uranium, or the sun—to boil water to produce steam. The steam turns a turbine connected to a generator to produce electricity.

[4] Production of oil shale requires the heating of rock containing solid organic matter to between 650 and 1000 degrees Fahrenheit and injecting water into the formation to stimulate the oil to flow. To date, there has been no commercial production of oil shale resources, in part, because the energy requirements to heat the rock and the water needed to stimulate the flow of oil make the process too costly to compete with other sources of oil.

[5] The Pacific Institute conducts interdisciplinary research and partners with stakeholders to produce solutions that advance environmental protection, economic development, and social equity in California, nationally, and internationally.

[6] GAO-10-23.

[7] The Energy Policy Act of 2005 created a Renewable Fuel Standard that generally required gasoline and diesel in the United States to contain renewable fuels, such as ethanol and biodiesel. Pub. L. No. 109-58 § 1501 (2005). The Energy Independence and Security Act of 2007 expanded the Renewable Fuel Standard. Pub. L. No. 110-140 § 201 (2007).

[8] GAO-10-116.

[9] GAO-11-35.

[10] GAO-12-156.

[11] According to federal estimates, more than 90 percent of all produced water is managed through underground injection.

[12] GAO-11-225.

[13] GAO-10-116.

[14] GAO-10-23.

[15] GAO-12-156.

[16] GAO-11-225.

[17] GAO-10-23.

[18] GAO-10-23. GAO-10-116.

[19] GAO-10-116.

[20] GAO-11-225.

[21] GAO-11-225.

[22] GAO-10-23.

[23] GAO-11-35.

[24] An aquifer is a geologic formation or structure that stores and/or transmits water, such as to wells and springs. Aquifer recharge is the process by which water is added to an aquifer, such as through rainfall that seeps into the ground.

[25] GAO-10-116.

[26] Carbon capture and sequestration technologies separate and capture carbon dioxide from other gases produced when combusting or gasifying coal, compress it, then transport it to underground geologic formations where it is injected for long-term storage.

[27] In 2010, EPA promulgated a rule establishing minimum requirements under the Safe Drinking Water Act for the underground injection of carbon dioxide for the purpose of geologic sequestration. According to EPA officials, they are in the process of implementing the rule, and they believe the rule is fully protective of underground sources of drinking water.

[28] GAO-10-116.

[29] GAO-10-116.

[30] We reported on the need for downscaled climate information in *Climate Change Adaptation: Strategic Federal Planning Could Help Government Officials Make More Informed Decisions*, GAO-10-113 (Washington, D.C.: Oct. 7, 2009).

[31] GAO-10-23.

[32] GAO-11-35.

[33] GAO-10-23.

[34] GAO-11-35.

[35] DOE, *Energy Demands on Water Resources: Report to Congress on the Interdependency of Energy and Water* (December 2006).

[36] In many states, public utility commissions are responsible for approving the rates (or prices) electric utilities charge their customers and ensuring they are reasonable. As part of approving rates, these commissions approve utility investments and, as a result, may consider whether specific technologies are reasonable.

[37] The Office of Science is the lead federal agency supporting fundamental scientific research for energy and the nation's largest supporter of basic research in the physical sciences.

[38] Energy Policy Act of 2005, Pub. L. No. 109-58, §§ 971, 979, 119 Stat. 594, 905 (codified at 42 U.S.C. §§ 16311, 16319 (2006)). Specifically, the act directs the Secretary to carry out a program of research, development, demonstration, and commercial application to (1) address energy-related issues associated with the provision of adequate water supplies, optimal management, and efficient use of water; (2) address water-related issues associated with the provision of adequate supplies, optimal management, and efficient use of energy; and (3) assess the effectiveness of existing programs within the department and other federal agencies to address these energy and water related issues. The program is to include, among other things, planning, analysis, and modeling of energy and water supply and demand.

[39] The Office of Science may be able to enter into inter- and intra-agency agreements, under authorities like the Economy Act, with other federal offices to assist in carrying out the provision to the extent those offices are able to provide relevant assistance. *See* 31 U.S.C. § 1535 (2006).

[40] H.R. 5827, 112th Cong. (2012).

[41] H.R. 5826, 112th Cong. (2012).

[42] Shale oil and shale gas refers to product that is trapped within underground shale formations; these fine-grain sedimentary rocks can be rich sources of oil and natural gas. Unlike oil shale, shale oil and shale gas do not require heating for resource extraction.

[43] GAO-11-35.

[44] The report looked at a variety of scenarios, ranging from 30- to 90-percent renewable generation in 2050. The report focused on 80-percent renewable generation because the laboratory's analyses concluded that renewable energy generation from technologies that are commercially available today, in combination with a more flexible electric system, would be technically adequate to supply 80 percent of projected U.S. electricity generation in 2050, albeit at increased electricity prices compared with today's baseline. Nearly 50 percent of the renewable electricity in this 80-percent scenario consists of variable wind and solar photovoltaic generation. (National Renewable Energy Laboratory, *Renewable Electricity Futures Study: Exploration of High-Penetration Renewable Electricity Futures*, vol. 1 (2012).)

[45] Solar photovoltaic panels and wind turbines use small amounts of water for panel and blade washing, respectively.

[46] Concentrating solar power plants operate by using mirrors to reflect and concentrate sunlight onto receivers that collect the solar energy and convert it to heat. This thermal energy is used to produce electricity via a steam turbine or heat engine driving a generator to produce electricity.

[47] According to DOE officials, concentrating solar power plants are generally being built with dry cooling systems in the Southwest to minimize water use. However, according to a 2009

DOE report to Congress, while dry cooling can eliminate over 90 percent of the water consumed by wet-cooled concentrating solar power plants, wet cooling is preferred to minimize cost and maximize efficiency.

[48] GAO-11-225.

[49] Ozonation is a water treatment process that destroys bacteria and other microorganisms through the infusion of ozone, a gas produced by subjecting oxygen molecules to high electrical voltage. Membrane filtration uses pressure to force untreated water through a semipermeable membrane, thereby filtering out bacteria and other microorganisms, particulate material, and natural organic matter.

[50] S uch standards may include drinking water standards set under the Safe Drinking Water Act or standards for the treatment of wastewater discharged from municipal wastewater treatment plants under the Clean Water Act.

End Note for Appendix I

[1] The Pacific Institute conducts interdisciplinary research and partners with stakeholders to produce solutions that advance environmental protection, economic development, and social equity in California, nationally, and internationally.

In: The Energy-Water Nexus: Key Issues ISBN: 978-1-62417-753-8
Editors: Tucker Webb and Ellis Porter © 2013 Nova Science Publishers, Inc.

Chapter 2

ENERGY'S WATER DEMAND: TRENDS, VULNERABILITIES, AND MANAGEMENT*

Nicole T. Carter

SUMMARY

The energy choices before Congress represent vastly different demands on domestic freshwater because water is used in varying amounts in most aspects of the energy sector. Transitions in the energy sector, such as the pursuit of greater energy independence and security, produce changes in how much and where the energy sector uses water. The energy sector is the fastest-growing water consumer in the United States, in part because of energy policies. Whether the federal government addresses the energy sector's rising water demand, and if so how, is one of the many energy decisions that may be considered by the 112th Congress.

Much of the growth in the energy sector's water demand is concentrated in regions with already intense competition over water. Whether the energy sector exacerbates or alleviates future water tensions is influenced by near-term energy policy and investment decisions. These decisions also may determine whether water will limit or harm U.S. capability to reliably meet the nation's energy demand. Part of the policy

* This is an edited, reformatted and augmented version of Congressional Research Service, Publication No. R41507, dated January 5, 2011.

issue for Congress is identifying the extent of the federal role in responding to the energy sector's water use. Currently, the energy industry and states have the most responsibility for managing and meeting the energy sector's water demand.

The energy sector's water consumption is projected to rise 50% from 2005 to 2030. Projections attribute to the energy sector 85% of the growth in domestic water consumption between 2005 and 2030. The drivers of the energy sector's increasing water use are rising energy demand, greater development of domestic energy, and shifts to more water-intense energy sources and technologies. The more water used by the energy sector, the more vulnerable the energy system is to water availability. Climate change alterations of historic water patterns may exacerbate this vulnerability in some regions. While the energy sector's water demand is anticipated to rise across the United States, the West is likely to experience some of the more significant constraints in meeting this demand. Examples of regional water use concerns related to energy are shale gas production using hydraulic fracturing in many regions across the nation, some solar electricity generation in the Southwest, and current biofuel feedstock production in the High Plains.

The 112th Congress may see the issue of energy's water demand in a variety of contexts, including oversight and legislation on energy, infrastructure, environment, agriculture, public lands, climate, research, and water. Approaches for addressing energy's water demand range from maintaining the current approach to taking a range of actions. One set of available actions includes those that attempt to minimize the growth in energy's freshwater use (e.g., through promotion of water-efficient energy alternatives and energy demand management), which could be accomplished through changes to broad policies or legislation targeted at water use. Many of the possible actions to decrease water use come with energy cost, generation, and reliability penalties. Another set of actions includes measures that facilitate access to water for the energy sector. While water allocations and permits generally are a state responsibility, limited federal actions to provide water to the energy sector are possible (e.g., access to surplus water at federal reservoirs). An additional set of actions encompasses investments in data and research to inform decision making and to expand water-efficient energy technology choices. These approaches represent different potential roles and expenses for government, the energy sector, and energy consumers. Legislation in the 111th Congress proposed a variety of actions, including provisions in H.R. 469, H.R. 2454, H.R. 3598, H.R. 1145, S. 1462, S. 1733, S. 3396, and the Secure Water Act of 2009, Subtitle IV of P.L. 111-11 (H.R. 146). A significant challenge to a federal response to the energy sector's water demand is that the available options are not equally needed, attractive, or feasible across the United States.

ENERGY'S WATER USE: A POLICY INTRODUCTION

The nation's energy choices embody many tradeoffs. Water use is one of those tradeoffs. The energy sector is the fastest-growing water consumer in the United States. Projections attribute to the energy sector 85% of the growth in domestic water consumption between 2005 and 2030. This projected growth derives from anticipated demand for more energy, increased development of domestic energy sources, and greater use of water-intense energy alternatives. Much of the energy sector's growing water demand is concentrated in water-constrained regions. Affordable water supplies are limited, and competition for water is becoming more intense. Whether the energy sector helps exacerbate or alleviate future water tensions is influenced by current energy policy and investment choices. These choices also may determine whether water limits or harms the nation's ability to reliably meet energy demand. Water limitations may hinder some water-dependent energy activities in specific locations.

Water already plays a significant role in the energy sector, and water use by the energy sector already shapes national water use. For example, more than 80% of U.S. electricity is generated at thermoelectric facilities. With few exceptions, these thermoelectric power plants are cooled with water.[1] In 2005, withdrawal of water for cooling represented 44% of water withdrawn nationally,[2] and 6% of the water consumed nationally.[3] The more water used by the energy sector, the more vulnerable energy production and reliability are to competition with other water uses and water constraints such as droughts. Climate change impacts on water supplies in some regions may exacerbate this vulnerability. Water availability can affect both existing and new energy activities, as well as all those economic activities that depend on the fuels and electricity produced.

The energy sector is changing. Paths chosen and capital investments made in the near term are likely to establish long-term trajectories for energy's water use. Trends indicate that energy's changing water use has national and regional significance for water consumption. A question for Congress is: what is the appropriate federal role in responding to energy's water demand? In the aggregate, current federal energy policies contribute to energy's rising water demand, while energy interests and the state and local governments are responsible for managing and meeting water demands and resolving competition over water resources.

Questions for Congress include *who* is the most appropriate entity to respond to energy's growing water demand and water vulnerability and *how* to respond. At present, little direct federal action is aimed at managing the energy sector's water demand; instead, the current division of responsibilities relies on energy interests and state and local governments to meet and manage energy's water demand and resolve energy-water conflicts. The role of federal policies in contributing to rising water demand is bringing into question the future federal role in this policy arena. Local or regional competition for water with existing users is often what makes energy's water demand significant; at the same time, the regional and local scales of water resources availability and management complicate many federal water-related actions.

Options for managing and meeting energy's water demand range from maintaining the current approach, with little federal action targeted at managing energy's water demand, to taking a variety of federal actions. One option is to minimize growth in energy's freshwater use. This could be accomplished through changes to broad policies (e.g., energy demand management) or legislation specifically targeted at water use (e.g., promotion of water-efficient energy alternatives). Another option is to improve the energy sector's access to water. Access is generally a responsibility of the state, but some limited federal actions are possible. An additional option is investing in data and research to inform decision-making and expand water-efficient energy technologies. These alternative policy approaches, which are not mutually exclusive, represent different potential roles and costs for the energy sector; energy consumers; and federal, state, and local governments. Legislation in the 111th Congress proposed many of the above options; examples include H.R. 469, H.R. 2454, H.R. 3598, H.R. 1145, S. 1462, S. 1733, S. 3396, and Subtitle IV of P.L. 111-11 (H.R. 146), the Secure Water Act of 2009.

During the 112th Congress, energy's water use may arise in a variety of contexts, including during consideration of energy, agriculture, public land, and water legislation and oversight. While the water tradeoffs of energy choices may be raised in a variety of contexts, they are unlikely to be the focus of energy debates. Instead, the priority on and investments toward different policy goals—low-cost reliable energy, energy independence and security, climate change mitigation, and job creation—are likely to be more significant drivers in congressional energy deliberations.

Scope and Structure of This Report

This report focuses on the factors shaping the energy sector's water demand, how that demand fits into national water consumption, and options for managing and meeting water demand. The report first lays out the trends shaping energy's water use; second, it discusses energy's vulnerability to water constraints; and third, it discusses projections of energy's water use. It then explores three regional examples of energy's water use: shale gas in Texas, solar energy in the Southwest, and biofuels in the High Plains. Finally, it discusses policy options and legislative approaches for managing energy's water use. Several appendixes provide more detailed information on specific technologies, fuels, and trends.

The report does not discuss in detail the energy sector's water quality impacts, although they represent their own challenges, as shown by concerns over the water quality effects of hydraulic fracturing, mountaintop mining,[4] and the Deepwater Horizon oil spill.[5]

Energy use by the water sector also is not discussed in this report, although water conservation is one of many available means for reducing energy demand.

The Federal Role in Water Allocation, Management, and Planning

The United States is endowed with considerable water resources; even in the drier West there are some major rivers and substantial groundwater supplies. Responsibility for development, management, and allocation of the nation's water resources is spread among federal, state, local, tribal, and private interests.

Water Allocation

The states administer most water rights allocation, deciding how to distribute freshwater among users. The mechanisms for distribution differ for each state, with states in the East and West historically following two different systems of water law. Distribution becomes more contentious as local or regional demand and environmental needs outpace sustainable and affordable supplies. States in large part decide whether and how to adapt allocation mechanisms and institutions to meet changing demands and priorities, and whether to allow or facilitate the movement of water among alternate uses. However, the federal role in allocation increases as the ramifications of state water allocations are felt in other states or internationally, or when they run contrary to federal law, such as the Endangered Species Act or the Clean Water Act.

Rather than directly influencing state allocation laws, decisions, and institutions, the federal government usually uses indirect means to influence water use, such as programs for agricultural water conservation, investments in water augmentation research, and support for water resource planning.

Water Management

The federal government has been called upon to assist with and pay for a multitude of water resource development projects. Federal works range from improvements to facilitate navigation beginning in the earliest days of the nation, to more recent efforts to reduce flood damages and expand irrigation. In recent decades, the federal government also has regulated water quality, protected fish and wildlife, and facilitated some water supply augmentation. Criticism of the fractured nature of federal water policy and concerns about the efficiency of current water use have been recurrent themes for decades. Congress has not enacted comprehensive changes in federal water resources management or national water policy since enactment of the 1965 Water Resources Planning Act (P.L. 89-80; 42 U.S.C. § 1962). While many stakeholders call for better coordination and a clearer national "vision" for water management, Congress often has reacted to such proposals as attempts to exert federal control over state and local matters or as attempts to concentrate power in the executive branch. Instead, Congress has enacted numerous incremental changes, agency by agency, statute by statute. Both the executive and judicial branches have responded to these changes and, over time, have developed policy and planning mechanisms on an ad hoc basis. When coordination of federal activity has occurred, it has been driven largely by pending crises, such as potential threatened or endangered species listings, droughts, floods, and hurricanes; and by local or regional initiatives. Concern about water supply, however, has bolstered recent interest in legislation to establish a national water commission and strategy.

Water Planning

Energy's rising water demand and the potential for climate change to decrease water supply availability and reliability in some regions are part of a recent interest in water planning.

Consensus, however, does not exist about the utility of or proper federal role in planning. Following the 1965 Water Resources Planning Act, the federal government supported federal, state, and river basin planning. By the late 1970s, this planning was both positively received and criticized for its costs and usefulness. The challenge of water planning is summed up in a 1973 report:

> A persistent tendency of water resources planning has been the issuance of single valued projections of water use into the future under a continuation of present policies, leading to astronomical estimates of future water requirements.... The amount of water that is actually used in the future will depend in large measure on public policies that are adopted. The National Water Commission is convinced that there are few water "requirements."... But there are "demands" for water and water-related services that are affected by a whole host of other factors and policy decisions, some in fields far removed from what is generally considered to be water policy. (National Water Commission, *Water Policies for the Future: Final Report to the President and to the Congress of the United States* (Washington: GPO, 1973), p. 2)

In the late 1970s and early 1980s, federal funding for state planning declined, and federal involvement shrank with the defunding of the executive-level Water Resources Council and most federal river basin commissions. Some water resources stakeholders continue to view federal involvement in planning as infringing on state primacy, while other stakeholders support greater federal involvement in watershed, multi-objective, or integrated planning efforts. Some states, such as California, Texas, and Florida, have undertaken their own planning efforts.

ENERGY TRENDS SHAPE WATER DEMAND

Trends in national energy investments, domestic energy use, population, and climate change impacts and responses can affect *how much* and *where* the energy sector uses water. Increased emphasis on domestic energy production and efforts to meet increasing energy demand are expected to increase freshwater use by the energy sector. A shift in the electricity sector away from traditional coal power plants may result in either more or less water consumption, depending on alternative fuels or electricity technologies.

Carbon capture and sequestration (CCS) by electric utilities has the potential to consume significant quantities of water.[6] Actions such as substituting wind for thermoelectric electricity generation could potentially reduce energy's water demand, but may raise other challenges for energy reliability, dispatchability, and transmission. Other impacts, such as the movement of irrigated agriculture from food crops to energy crops, raise other concerns. These and other examples of energy trends and their effects on energy's water use are summarized in Table 1. Appendix A has a more extensive list of trends.

Table 1. Energy Trends Produce Water Use Trends

Energy Trend	Resulting Trend in Energy's Water Use
Shift from foreign oil to biofuels	Increases energy's water consumption if domestic agricultural irrigation water (and other inputs) is needed for fuel production.
Shift to shale gas	Natural gas development using hydraulic fracturing may raise water quantity concerns if well development is geographically concentrated in areas with water constraints. However, natural gas from fracturing consumes less U.S. freshwater than domestic ethanol or onshore oil.
Growth in domestic electricity demand	More water used for electricity generation; how much more depends on how the electricity is produced (e.g., smaller quantities needed if electricity demands are met with wind and photovoltaic solar, larger quantities if met with fossil fuels or certain renewable sources).
Shift to renewable electricity	Concentrating solar power technologies can use more water to produce electricity than coal or natural gas; these solar facilities are likely to be concentrated in water-constrained areas. Technologies are available to reduce this water use. Other renewable technologies, such as photovoltaic solar and wind, use little water.
Use of carbon mitigation measures	Carbon capture and sequestration may double water consumption for fossil fuel electric generation.

Source: CRS.

Whether and how much the energy sector's water demand grows in the next decades will be significantly influenced by whether energy demand increases. Projections of the size and mix of the future energy portfolio vary widely. These projections are highly uncertain and are sensitive to many factors, including market and economic conditions, energy and agricultural policies, resource availability, technology developments, and environmental regulations. By association, projections of energy's water demand also are highly uncertain.

The Energy Information Administration (EIA) in the Department of Energy (DOE) projects that the United States will consume 22% more electricity and 12% more liquid fuel in 2030 than in 2010.[7] Population growth and increased electricity use per capita are some drivers of increasing demand.

Significant shifts to more water-intense electricity generation (e.g., concentrating solar power facilities using evaporative cooling) or more water-intense fuels (e.g., oil shale) could increase energy's water demand in locations with these energy resources.[8]

The significance of energy's water demand depends in part on local conditions—how much water is locally available and what its alternative uses would be.

The most growth in freshwater consumption by the energy sector is expected in the Southwest, the Northwest, and the High Plains—that is, regions already experiencing intense competition over water and disputes over river and aquifer management.

Transportation's Water Consumption Is Increasing

The transportation sector offers an example of how energy trends affect water use. Water consumption by transportation fuels is anticipated to increase between 2005 and 2030; this increase is shaped by multiple trends.

As shown in the figure below, an increase in miles driven and the increasing water-intensity of fuels, as a result of irrigated biofuels (i.e., biofuels derived from irrigated feedstock), overwhelms the water gains from improving vehicle fuel efficiency. (Appendix B has more information on water consumed for a variety of transportation fuels.)

For light-duty vehicles (LDVs), the Energy Information Administration (EIA) projects a 50% increase in miles traveled, while the water-intensity of those miles is projected to rise from 40 gallons per 100 miles traveled to almost 90 gallons, according to researchers. Consequently, water consumption for LDV travel is roughly projected to triple from 2005 to 2030.

Water consumption for LDV travel increased 50% between 2005 and 2010.

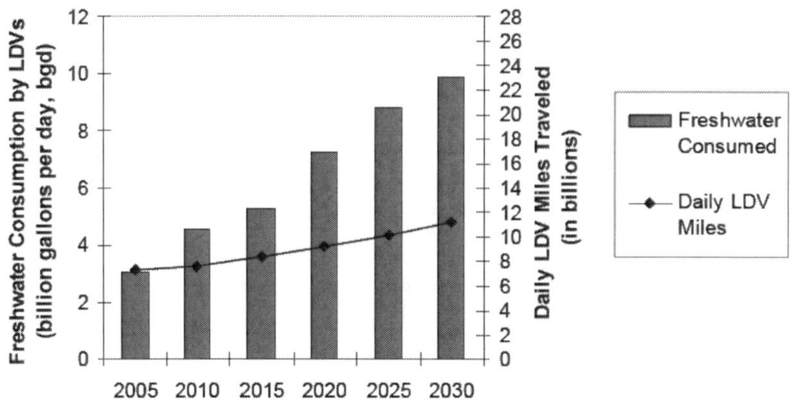

Sources: C. W. King, M. E. Webber, and I. J. Duncan, "The Water Needs for LDV Transportation in the United States," *Energy Policy*, vol. 38 (2010), pp. 1157-1167; EIA, *Annual Energy Outlook 2008*, Table 7, Transportation Sector Key Indicators and Delivered Energy Consumption, Line 15. The King 2010 article used 2008 EIA data.

ENERGY SECTOR'S VULNERABILITY TO WATER CONSTRAINTS

The more freshwater used by the energy sector, the more the sector is vulnerable to water constraints. However, as described above and in Appendix A, major energy trends are pushing the sector to become more water-intensive. Water availability problems, especially regional drought and low streamflow, can pose a risk to energy production and reliability. Electricity generation is particularly sensitive to low-flow conditions. More than 80% of U.S. electricity is generated at thermoelectric facilities that depend on access to cooling water. Low-flow conditions and water scarcity may constrain water-intense alternatives for thermoelectric cooling in counties across the country. Additional ways that water can constrain energy include a possible decrease in hydroelectric generation during drought. Bioenergy yields may be reduced by low precipitation, droughts, heat waves, or floods. Energy extraction, like coal mining, may be scaled back to avoid water quality impairments exacerbated by low water conditions. While water constraints are often perceived as an issue for the western United States, an increasing number of water bodies in the East are experiencing diminished stream flows.[9] While multiple examples exist of water availability affecting siting and operations of thermoelectric facilities from New York to Arizona,[10] generally there are ways to reduce the use of water and the risk posed by water constraints.

Water supplies often are most constrained during summer, when the energy sector's water use is at its height in many regions. Approximately 24 of the nation's 104 nuclear reactors are situated in drought-prone regions.[11] The Nuclear Regulatory Commission sets minimum source water elevation levels for each plant, so that a plant does not operate when source water levels drop below plant cooling water intakes. If cooling water sources fall below the established minimum water level, or if the maximum thermal threshold for the discharge of cooling water cannot be met, a facility is required to power down or go offline. A commonly cited example of this occurred on August 16, 2007, when a nuclear reactor owned by the Tennessee Valley Authority (TVA) at the Browns Ferry Nuclear Power Plant in Alabama shut down for a day. Its cooling water discharge exceeded temperature regulations that protect the environment and wildlife of the Tennessee River. In the summer of 2010, the same plant cut its electricity production to 45% of capacity when the cooling water temperature again exceeded discharge regulations.[12] The reduced generation resulted in $50 million in higher cost to customers. TVA subsequently initiated a $160 million upgrade and expansion of its cooling system to avoid similar cooling constraints in the future.[13]

Water Constraints on Thermoelectric Cooling

More than 80% of U.S. electricity is generated at thermoelectric facilities. Thermoelectric facilities generally can be used to produce power as needed, according to consumer demand and fuel supply. This responsiveness to demand makes electricity from these thermoelectric facilities particularly attractive.

Thermoelectric facilities can be fueled by a variety of fuels; coal, nuclear, and natural gas are the most common. Renewable sources such as concentrating solar power (CSP), geothermal, and renewable biomass also use a thermoelectric steam cycle.

Thermoelectric power plants use fuel to produce heat to generate steam, which turns a turbine connected to a generator that produces electricity. Cooling is required to condense the steam back into boiler feed water, so the process can be repeated. With few exceptions, water is used to cool existing U.S. thermoelectric power plants. Thermoelectric cooling represents 44% of the freshwater withdrawn nationally, and roughly 6% of the water consumed nationally.

The cooling options available for thermoelectric plants vary in their water withdrawal and consumption. Water withdrawal is the volume of water removed from a water source. Consumption is the volume lost, that is, no longer available for use. Excessive withdrawals can harm aquatic ecosystems, while excessive consumption depletes the water available for other uses. The two common cooling methods are once-through cooling and evaporative cooling. Once-through cooling pulls large quantities of water off a water body, discharges the power plant's waste heat into the water (which typically raises its temperature 10° to 20°F), then returns the majority of the withdrawn water.

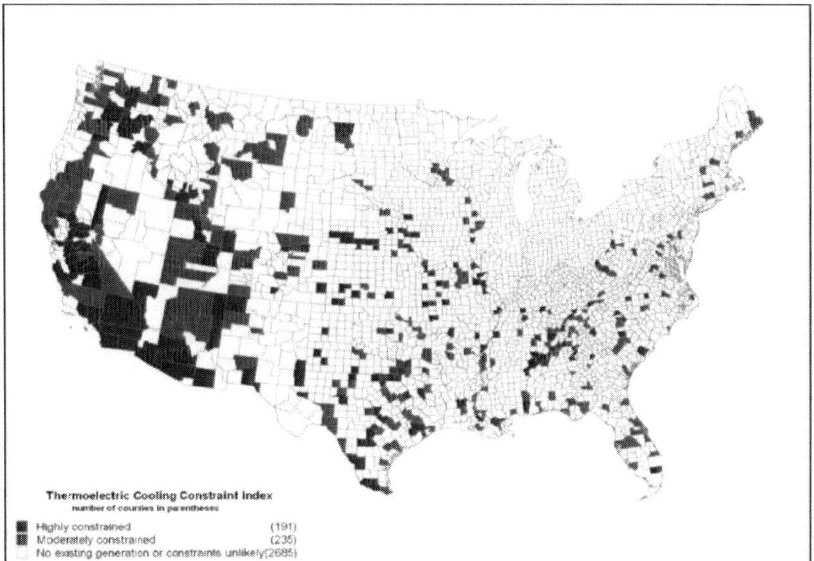

Source: EPRI, A Survey of Water Use and Sustainability in the United States with a Focus on Power Generation, Topical Report, Nov. 2003. EPRI's analysis did not include Alaska and Hawaii.

Once-through cooling, while largely a non-consumptive water use, requires that water be continuously available for power plant operations. This reduces the ability for this water to be put toward other water uses and can make cooling operations vulnerable to low streamflows. Evaporative cooling withdraws much smaller volumes of water for use in a cooling tower or reservoir, where waste heat is dissipated by evaporating the cooling water.

Evaporative cooling consumes water. Many power plants operating in the East use once-through cooling, while the majority in the West use evaporative cooling, although some coastal facilities use saline water for once-through cooling. In general, older thermoelectric plants use once-through cooling. The withdrawal's effect on the ecology and quality of the water body (e.g., elevated temperature and chemicals of the discharged cooling water) in once-through cooling have resulted in newer power plants generally using evaporative cooling. The figure below was produced by Electric Power Research Institute (EPRI) in 2003. It projects the counties where thermoelectric cooling may be constrained as the result of water availability in 2025; this figure was created assuming historic water availability; that is, it did not account for potential climate change effects on water supply or electricity demand.

Climate Change Could Increase Energy's Freshwater Vulnerability

Snowpack, precipitation, and runoff are strongly related to climate. Climate change researchers predict both water quantity and timing changes. That is, the research indicates more precipitation in the form of rain and less in the form of snow, and changes to seasonal water availability in some areas (e.g., low-flows during dry seasons). Additionally, climate models predict more frequent floods and droughts. These changes present challenges for hydroelectric dam operation.[14]

Changes in the availability and temperature of water resources also may affect operations of power plants that require water for cooling and that have thermal discharge limitations for cooling water. Climate change also may increase the demand for air conditioning, the electricity it consumes, and the water used to produce the electricity. The decreased runoff anticipated in the West, Southwest, California, and Pacific Northwest[15] would decrease the amount of water available for all uses, including the energy sector.

That is, the water resource impacts of a changing climate would likely exacerbate already projected thermoelectric cooling constraints. The energy sector also is vulnerable to potential increased flood and storm hazards associated with climate change. An example is the disrupting effects of floods on fuel transport.[16]

How Much Water Does Energy Demand?

How much water will the energy sector use in the future? In part, interest in answering this question is rooted in other questions of national significance, such as: Will water limit U.S. capacity to meet the nation's energy demand? Will water constrain the transition to greater energy independence? Will water hamper adoption of some renewable energy alternatives?

Quantification of energy's water demand and its significance is limited by significant gaps in available data and analyses. Water has no federal data agency comparable to the Energy Information Administration that projects alternative demand scenarios.[17] There is no authoritative government source to cite for the level of water use by the energy sector or for projections of how that use may change in future decades. For example, there are no forecasts that use multiple scenarios to identify sensitivity of water demand to multiple factors and policies, or that analyze energy's water use and water vulnerability in the context of factors significant to energy choices and policies, such as energy and transmission costs, emissions, and reliability.[18]

For over 50 years, the U.S. Geological Survey (USGS) of the Department of the Interior has collected and published water use data every five years; however, the agency stopped collecting water consumption data after its 1995 survey due to funding constraints and data reliability problems. (USGS continues to collect water withdrawal data.) The 1995 USGS data are the basis of most projections of future water consumption in the United States.[19] In the Secure Water Act of 2009, Congress authorized the USGS to perform a water use and availability assessment that includes water use trends in the energy sector; however, to date the agency had received no congressional appropriations for this program. Every 10 years the Forest Service forecasts water resources trends based largely on extrapolations of the USGS data and using the USGS data categories.[20] The energy sector falls into a number of USGS water use categories, and it is impossible to disaggregate the categories to determine a value for the energy sector's water use. (Except for thermoelectric water withdrawals, the USGS water use data and Forest Service projections do not break out energy water use from other agricultural and industrial water uses.) Because of these limitations, the analysis herein relies on the most comprehensive projections available on the energy sector's water consumption, with the main source being a 2010 article by Deborah Elcock, an Argonne National Laboratory researcher, based on an updated and refined analysis from a report published by the lab in 2008.[21]

Although currently available data are limited, there are prospects for improved data and analysis in the future. The Secretary of the Interior announced in October 2010 that the USGS is undertaking a Colorado River Basin Geographic focus study as part of the department's WaterSMART Water Availability and Use Assessments initiative; this focus study may eventually comprise a component of the USGS water use and availability assessment Congress authorized in 2009, which would represent the first national water census since 1978.[22] Additionally, regional efforts may inform future decision-making. For instance, the Western Governors' Association initiated in 2010 an energy-water nexus project, as part of its renewable energy transmission expansion effort, which includes a water availability assessment.[23] The assessment is looking into projected water demands for large river basins and aquifer systems in the West and is expected to consider drought and potential climate change implications on the availability of river flows and water supply for energy development in the West. The effort is anticipated to conclude with policy recommendations available in late 2012, and is funded primarily by DOE.

Energy Leads Projections of Increasing U.S. Water Consumption

Energy's Water Demand May Increase 50% from 2005 to 2030

During the 1980s and 1990s, as a result of improved water efficiencies, U.S. water consumption remained below the historic high level of 101 billion gallons per day (bgd) estimated for 1980.[24] Estimates of recent consumption, however, put U.S. freshwater consumption above the previous high and predict it will increase further in the next decades. (See Figure 1.) The projected rise is dominated by energy's water use. Nationally, energy's water consumption exceeds municipal and industrial use; it is currently second only to agriculture.

By modeling current energy and water trends, available projections predict that water consumption by 2030 will increase by 7% above the level consumed in 2005, as shown in Figure 1. Eighty-five percent of this growth is attributed to the energy sector, and an increase has already been observed between 2005 and 2010, with greater energy sector water use for irrigated biofuels.[25] (See Appendix B.) Energy's water consumption (excluding hydropower) in 2005 was roughly 12 bgd, as shown in Figure 2; it is estimated to increase to 18 bgd by 2030. As shown by the data summarized in Figure 2, roughly 60% of the anticipated expansion is associated with bioenergy, and 40% is associated with thermoelectric cooling and fossil fuel mining, production, and processing. The analyses behind these projections do not capture the effects on water consumption of various fuel and generation technology changes in the electricity sector and potential adoption of CCS, nor do they include hydropower's water consumption.

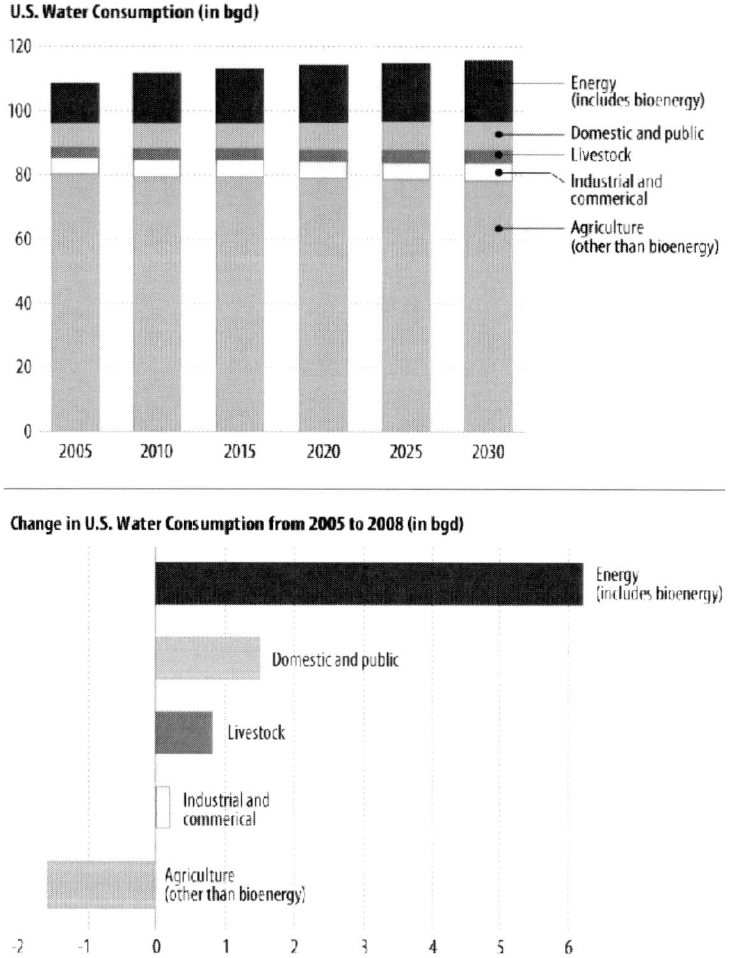

Source: CRS, using data from D. Elcock, "Future U.S. Water Consumption: The Role of Energy Production," *Journal of the American Water Resources Association*, vol. 46, no. 3 (June 2010), pp. 447-480.

Figure 1. Projection of U.S.Water Consumption.

Although the total national increase in Figure 1 may not appear large (7% over 25 years), multiple factors make this a significant increase for the water sector. First, growth in water consumption between 2005 and 2010 has put current water consumption above the previous high set in 1980. Second, water in many basins is largely allocated (or even over-allocated), with the water being delivered under legally binding agreements or withdrawn under issued permits. This means that making water available to the energy sector would likely decrease water use in another sector, such as agriculture. Third, Figure 1 represents national freshwater use; local proportions and increases in water demand from energy may be significantly higher.

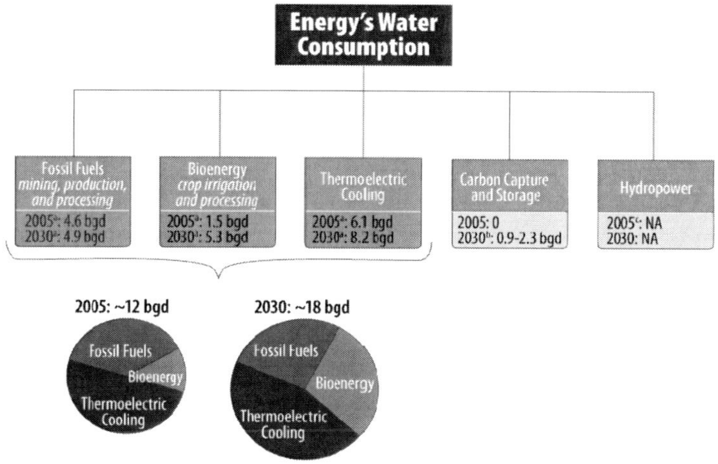

Source: CRS, using data sources noted below. Notes: NA = not available.

a D. Elcock, "Future U.S. Water Consumption: The Role of Energy Production," *Journal of the American Water Resources Association*, vol. 46, no. 3 (June 2010), pp. 447-480.

b NETL, *Estimating Freshwater Needs to Meet Future Thermoelectric Generation Requirements (2009 Update)*, p. 64.

c A 2003 National Renewable Energy laboratory (NREL) report estimated that evaporation at reservoirs at the 120 largest existing U.S. hydroelectric facilities represented 9 bgd (see P. Torcellini et al., *Consumptive Water Use for U.S. Power Production* (NREL, 2003), available at http://www.nrel.gov/docs /fy04osti /35190.pdf., hereafter referred to as NREL 2003). This estimate is not shown in the figure above because these reservoirs are used for multiple purposes; therefore, evaporation at these reservoirs cannot be attributed to hydropower alone.

Figure 2. Energy's U.S. Freshwater Consumption, (in gallons per day, bgd).

To illustrate the role that energy's water demand plays at the state level, roughly 16% of the water diverted in Kansas in 2008 went to biofuels; another 16% was used for electric power generation, while roughly 9% was used for municipal purposes.[26] Fourth, in some regions, climate change is anticipated to decrease the quantity and reliability of water supplies. In particular, lower-altitude and drier areas are anticipated to become drier; some regions (e.g., the West) may experience more frequent or intense drought years punctuated by wet years.[27] Lower precipitation in a portion of the Great Plains, and reduction in temperature-vulnerable snowpack in the western near-coastal mountains, are some of the anticipated changes to water supply.[28] Therefore, energy's demand for more water use may coincide with a decreasing and less predictable water supply.

Projections are based on assumptions that are subject to change; in particular, management of energy demand and wide deployment of water-efficient energy options may significantly lower water use below projected levels. A DOE Energy Efficiency and Renewable Energy (EERE) study found that expanding the nation's electricity portfolio to 20% wind by 2030 would reduce water consumption by 1.2 bgd compared to expanding the current electricity mix. The saved water would be 41% from the Midwest/Great Plains, 29% from the West, 16% from the Southeast, and 14% from the Northeast.[29]

Alternatively, the projections in Figure 1 could be underestimates. Some factors that may augment energy's water consumption are not accounted for in these projections. More water could be consumed if carbon capture and sequestration is widely employed (see discussion below), or if more water-intense energy is produced. For example, significant expansion of electricity from evaporative-cooled concentrating solar power (CSP), or new hydropower reservoirs in areas with high evaporation rates, could drive up energy's water consumption. Similarly, a significant increase in oil shale development could increase energy's water use in regions with those deposits. In summary, projections based on extrapolations of current trends can illustrate one potential path for water consumption, but many factors, such as the role of changing technologies and policies, can result in actual water consumption varying significantly from projections.

Carbon Capture May Create a New Energy Water Demand

The majority of current U.S. electric generation is from fossil fuels. The electric generation mix in 2009 was 45% coal, 23% natural gas, 20% nuclear, and 7% hydroelectric, less than 4% non-hydroelectric renewable generation, and less than 2% other sources.[30] Carbon capture technologies in the fossil fuel industry may reduce carbon dioxide emission, but they come with investment and resource costs to manufacture and operate. Current carbon capture technologies consume energy and water.[31]

Water is used at two points in the carbon capture cycle: cooling water is required for capture and compression processes, and water generally is consumed for power plant cooling when generating power needed to perform CCS. A 2009 study by the National Energy Technology Laboratory found that by 2030, carbon capture and sequestration could increase water consumption for electric generation by anywhere from 0.9 bgd to 2.3 bgd, depending on the scenarios used for plant additions and for deploying carbon capture technologies.[32]

Hydroelectric Water Consumption Is Poorly Documented

Although most hydropower generation represents an in-stream water use, dams built to generate hydropower and for other purposes consume water by increasing evaporation above free-flowing river conditions. That is, more evaporation occurs at the reservoir behind the dam than at the river without the dam.

How much evaporation occurs at a reservoir with a hydroelectric generation depends on site, climate, and water conditions. CRS was unable to locate national data on existing or future projections for water consumption for hydroelectric generation.

The currently available water consumption data from the USGS do not include hydropower evaporation. The agency states: "Although the quantity of water evaporated in the actual generation of hydropower (consumptive use) is small, considerable depletion of the available water supply for hydroelectric power generation occurs as an indirect result of evaporation from reservoirs and repeated reuse of water within pumped-storage power facilities."[33]

As noted in Figure 2, a 2003 NREL report estimated evaporation at 120 reservoirs with hydroelectric facilities at 9 bgd.[34] Because these reservoirs generally serve multiple purposes, this evaporation cannot be attributed to hydropower alone and is not shown in the figure.

Moreover, many reservoirs enhance water availability for multiple functions by providing storage that regulates streamflow, at times minimizing the occurrence of naturally occurring low-flow levels.

These benefits, however, often come with harm to aquatic ecosystems, species, and floodplains. The NREL data illustrate that evaporation at facilities with hydropower can vary widely based on geography and climate.[35] Site-specific estimates of water consumption for new hydroelectric generation using new reservoirs (e.g., some pumped storage proposals), therefore, would be required to understand hydroelectric generation's impacts on water resources. That is, new hydropower reservoirs cannot be assumed to have minimal effect on water consumption.

Efficiency improvements or additions of hydropower generation at existing facilities, however, have the potential to increase electricity generation without increasing water consumption from evaporation.

REGIONAL SIGNIFICANCE OF ENERGY'S WATER DEMAND

Much of the anticipated growth in the energy sector's water demand is in water-constrained areas, potentially exacerbating competition and low flow condition for water during summer and droughts. That is, while energy's water demand is anticipated to rise across the United States, the West is likely to experience some of the more significant constraints and conflicts in meeting this demand. While local or regional competition for water is often what makes energy's water demand significant, the regional and local scales of water resources and how they are managed often complicate federal water-related actions.

To illustrate one aspect of how energy's growing water demand varies across the country, Figure 3 shows projections for expanded electricity generation in many water-constrained states (e.g., California, Texas, Arizona), primarily owing to rising electricity demand.

The following examples of regional water consumption concerns are discussed in more detail: shale gas production in Texas using hydraulic fracturing, solar electricity generation in the Southwest; and biofuel production in the High Plains.

Technologies and management practices exist for reducing water use in each of these energy activities; however, these options generally come with energy cost, generation, and reliability penalties.

Whether the benefits from the water savings of adopting these technologies and practices are viewed as outweighing these penalties often is a matter of perspective and is site specific. Moreover, little reliable and systematic data is available on the costs and benefits of adopting these measures in a variety of circumstances to better inform decision-making.

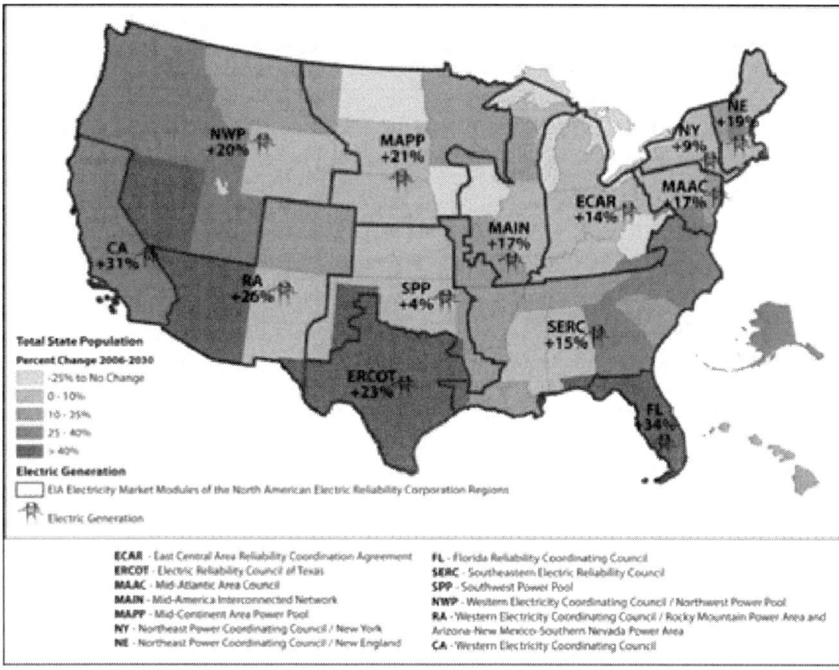

Source: EIA, *Supplemental Tables, Updated Annual Energy Outlook 2009 Reference Case with ARRA*, Wash., DC, April 2009, http://www.eia.doe.gov/oiaf /aeo/supplement/stimulus/suparra.htm); and U.S. Census Bureau, *U.S. Population Projections*, March 2004. EIA did not include Alaska and Hawaii.

Figure 3. Population and Electricity Generation Projections, 2010 to 2030.

Energy Development: Shale Gas in Texas

Shale gas development has expanded in the last decade, in part because of technological advances in horizontal drilling and hydraulic fracturing. Fracturing involves the pressurized injection of water-based fluids (water, sand, and chemical mixtures) into a well to fracture a rock formation so that natural gas is released. Shale gas formations occur in many areas of the United States.[36] Exploiting this resource is bringing the oil and gas industry into communities unaccustomed to energy development. Water use is one of a number of issues being raised with the expansion of unconventional gas development. While technological developments allow for greater extraction of gas, these technologies result in an increasing average amount of water used per well. That is, the longer the well, the more water is needed to drill and stimulate gas production. Freshwater, rather than saline water, is preferred for drilling and fracturing. Water use is concentrated in the early stages of well development, usually in the first few months. Once the well is producing, little or no water is required, unless refracturing is necessary. Much shale gas development is on private lands, and no government agency requires operators to report water use. Some data on water use per well are available, such as data from a DOE report in 2009 for four shale gas formations, which show water use ranging from 2.7 million gallons to 3.9 million gallons of freshwater per well.[37] Some limited data are available on the amount of water per unit of energy produced. Data from Chesapeake Energy shale gas wells indicate that drilling, fracturing, extraction, and processing use between 0.6 and 3.8 gallons of water per million British thermal units (MMBtu) produced, or 5 to 29 gallons per megawatt-hour (MWh), as fuel for a combined cycle natural gas facility (not including the power plant water use).[38] This puts shale gas at the lower end of water use for transportation fuels such as domestic onshore oil and irrigated biofuels, and below the water use for most coal and nuclear fuel production for electricity, according to a 2006 DOE report.[39]

The water use issue for natural gas is often one of concentration. That is, if many wells are being developed in a limited geographic area, the cumulative water needs of multiple drilling and fracturing operations may be locally significant and constraining to expansion of energy development. This may be particularly the case in areas with water constraints and competing water demands for domestic, agricultural, and thermoelectric use or areas where water for new gas operations represents a substantial expansion in water use.

Water quantity concerns have been raised for the Texas Barnett formation, particularly when drilling was using primarily fresh groundwater supplies and reached a peak of more than 3,000 new wells in 2008.[40] With the subsequent decline in gas prices, the number of new wells drilled annually has dropped to 1,000, and increasingly the water is coming from surface supplies purchased from municipal water utilities where available, such as in the Fort Worth area. Based on available projections, the maximum use for natural gas production annually in the affected Texas counties may reach 7.5 billion gallons (0.021 bgd) under a high drilling scenario; this would represent 1.15% of local water consumption.[41] While this shale gas use represents a modest share of local consumption, communities in this region of Texas have faced significant concerns about the sufficiency of available supplies during drought conditions for the region's growing population. This concern has brought scrutiny to all water uses, including shale gas. For most wells, the majority of the fluids injected into formation are not returned to the surface. In the Barnett formation, the water that is returned to the surface, known as flowback or produced water, is typically reinjected deep underground in a permitted disposal well. Texas A&M researcher C. J. Vavra estimates that more than half of the produced water could be reused in subsequent fracturing operations, and a quarter could be put to beneficial use.[42] These actions could reduce the impact of shale gas development on local water supplies but could raise other concerns. Legislation was considered by the 111th Congress to support research on this subject. For example, H.R. 469, the Produced Water Utilization Act of 2009, would have directed the Secretary of Energy to carry out a program to demonstrate technologies for environmentally sustainable use of energy-related produced waters from underground sources. Water quantity concerns related to shale gas development are often overshadowed by interest in the economic benefits of the gas development and other local concerns with development. These other concerns include the risk that fracturing-related activities may contaminate freshwater supplies, community disruption and air pollution from truck traffic related to gas development (e.g., truck transport of drilling equipment and materials and fracturing additives and freshwater on and wastewater off the well site), and short-term and long-term changes to the local landscape and community.

Although shale gas in Texas is discussed above, rising water demand for natural gas development is occurring or is anticipated in many other states and regions, such as the Northeast (Pennsylvania, New York), the South (e.g., Arkansas), and the upper Missouri River basin (Wyoming, South Dakota, North Dakota).

Renewable Energy Presents Regional Water Opportunities and Challenges

Water resource opportunities and challenges in meeting energy demand with renewable energy vary regionally.

- The Southwest has high-quality solar and geothermal resources, but water constraints and drought vulnerability favor more expensive, water-efficient use of these renewable resources.
- The Northwest generates most of its electricity from hydropower, but is diversifying to include more wind in light of environmental protections, increasing off-stream uses, and climate change effects on hydropower.

- The Great Plains may reduce water competition among the thermoelectric and agricultural sectors by
- exploiting wind power to produce electricity. Grid balancing and transmission of that electricity, especially to other regions, may pose constraints. Increases in irrigation to support bioenergy crops may tax stressed aquifers.
- The Southeast could reduce its dependence on coal and nuclear fuel and decrease its vulnerability to electricity interruptions during drought by producing electricity from biomass and photovoltaics.
- The Midwest could reduce competition between the thermoelectric and agricultural sectors and reduce the regional energy sector's low flow, drought, and flood vulnerability by exploiting its wind resources.

  This would require overcoming significant regulatory transmission issues. The Midwest is participating in the production of bioenergy crops, in particular corn for ethanol, raising water quality concerns, including that excess nutrients in agricultural runoff are feeding the growth of the "dead zone" in the Gulf of Mexico.
- The Northeast is not experiencing a regional issue with energy's growing water consumption; however, fracturing in shale gas formations is raising water quality concerns.
- Hawaii could transition to water-efficient power generation technologies to protect limited freshwater resources consumed in thermoelectric generation, which is dominated by oil-fueled power plants.
- Coastal regions, including Alaska, have the potential to develop offshore wind, tidal energy, waves, or ocean thermal gradients to reduce energy's onshore water use and land requirements. However, public opposition, transmission challenges, resource availability mismatched with demand, and natural hazards can pose challenges.

Source: CRS, with support from visiting researchers Ashlynn Stillwell and Kelly Twomey.

Electricity Generation: Solar in the Southwest

Another example of energy's regional water demand is concentrating solar power (CSP) in the Southwest. [43] The region has abundant solar resources, but the region's water constraints can influence the attractiveness and feasibility of different solar technologies and the suitability of specific sites. Most concentrating solar power (CSP) consists of ground-based arrays of mirrors that concentrate the sun's heat, which is used in a thermoelectric process to generate electricity. The two primary technologies are solar troughs and solar towers. Although CSP does not emit greenhouse gases, its use of a thermoelectric process can raise water concerns, particularly if evaporative cooling is employed (that is, if water is evaporated to dissipate waste heat). Similar water concerns would be raised if new fossil-fuel thermoelectric power plants were to be similarly located in the Southwest. For some Southwest counties with relatively low water use, large-scale deployment of CSP or another thermoelectric facility (even with water-efficient cooling technologies) could significantly increase the current demand for water in the county.[44]

Some solar developers are using cooling alternatives that require less water (or have been encouraged or required to do so as part of state permitting or federal approval of facilities on federal lands). These alternatives include dry or hybrid cooling or use of impaired waters for cooling (see Appendix C). These options generally come at a cost premium and with energy and cooling efficacy tradeoffs. Other solar developers have purchased water rights from willing sellers in states with active water markets. Still other solar developers are using solar technologies that require little or no water; these include photovoltaic solar, which uses panels of solar cells to convert sunlight directly into electricity, and dish engine CSP, which uses engines rather than a thermoelectric process[45] to produce electricity. (See Appendix C.) While these technologies are water-efficient, they have other constraints (e.g., cost, land use, dispatchability). In summary, freshwater constraints like those in the Southwest do not preclude solar development, but access to water shapes the technologies and costs of solar development.

Energy Crops: Biofuels in the High Plains[46]

An additional example of energy's regional water demand is biofuels in the High Plains. The expansion of biofuels use through incentives and consumption mandates is an example of how federal energy policies can affect national and local water use. The water quantity (and quality) used for biofuels is particularly sensitive to biofuel feedstock, use of irrigation, and local climate and soil conditions at the growing site. Average water consumption by the dominant U.S. biofuel, corn-based ethanol, significantly exceeds the water intensity of other U.S. transportation fuels if the corn feedstock is irrigated. (See Appendix B.) Irrigation of only a small amount of biofuel feedstock in areas without sufficient rainfall to support feedstock growth without irrigation has the potential to substantially increase national water consumption for transportation fuels. The High Plains—consisting of portions of Texas, New Mexico, Colorado, Kansas, Nebraska, Wyoming, and South Dakota—is one example of these low-rainfall areas. Much of the High Plains has faced water use issues for decades, such as the declining level of portions of the Ogallala aquifer since the mid-1960s.[47] Expansion of biofuels in this area is an additional demand exacerbating already competing water uses. For example, in Colorado in 2008, 16 times the annual quantity of treated water supplied to Fort Collins water customers was used to produce biofuels in the state, with the majority used in eastern Colorado.[48] Of particular concern to the High Plains is expanded biofuels production on new or marginal lands, which could lead to additional irrigation demand and increased nutrient application, causing both water quantity and quality concerns.[49] Even expansion of existing biofuel crops in current production, such as corn, raises water quality concerns with the possible increased application of fertilizers and pesticides necessary to increase yields (this is also of concern in regions such as the Midwest, where irrigation water concerns have been less significant).[50]

Recognition of water, land, and other issues related to biofuels, particularly irrigated corn- and soybean-based biofuels, has led to a search for feedstocks and other organisms that use fewer resources to produce.[51] Recent federal biofuels policies have attempted to assist this search by focusing on the development of a cellulosic biofuels industry.[52] Dedicated biomass crops, such as switchgrass, hybrid poplars, and hybrid willows are considered by many to be more desirable crops because they have a short rotation (regrow quickly after each harvest) and use fewer resources, such as water and fertilizers, than traditional field crop production. Despite potential environmental benefits, concerns persist about the additional use of fertilizers and water resources that could be required to increase the per-acre yields to economically feasible levels; for example, that cellulosic feedstocks may be irrigated to increase yields, even though irrigation may not be required.[53] Also, land use pressure for expanded production also applies to cellulosic biomass feedstock, possibly creating direct competition with current land conservation programs and replicating the concerns of traditional biofuel feedstock stated above. Despite federal incentives, technological and economic hurdles continue to prevent the cellulosic biofuels industry from developing to commercial scale production.[54]

MEETING AND MANAGING ENERGY'S WATER DEMAND: POLICY OPTIONS

The previous sections described how energy's demand for water is increasing and provided some regional examples. This section discusses options for meeting and managing that demand. Historically the energy sector and the states have determined how water is used in the energy sector, but the significant role that current federal policies are playing in driving up energy's water demand is raising questions about the federal role in meeting and managing that demand. These questions include:

- Are states being unfairly burdened with the responsibility of increased water use and competition resulting from federal energy policies, or is this part of the responsibility that comes with state primacy in water allocation?
- Who is responsible for the vulnerability of the nation's energy system to water availability?

Congress is faced with deciding not only whether, and if so how, to alter current policies to respond to energy's water demand, but also who is the most appropriate entity to respond to energy's growing water demand. Currently little direct federal action is aimed at managing the energy sector's water demand, although federal policies at times have significant indirect influence on this demand. Instead, present roles rely on the energy industry and the states and local governments to manage water constraints and to resolve energy-water conflicts.

The issue of the energy sector's water use may arise during the 112th Congress in a variety of contexts. Support or opposition for legislation affecting energy's water demand may be influenced by opinions about the proper federal role in water allocation and planning, as well as concerns about the cost of actions and who is responsible for those costs. Positions on the larger energy and climate debate and other factors may also be important. Federal responses to energy's water use are complicated by the wide-ranging and place-based nature of the issue, the variety of actors involved, the costs and other tradeoffs involved, and the existing institutions and divisions of responsibilities for water and energy.

If increased federal action to meet and manage energy's water demand is deemed appropriate, possible actions fall under a few broad options. Attempts can be made to minimize the growth in energy's freshwater use by adopting general energy policies that are less water intense and more sensitive to water constraints, or by specifically promoting activities that reduce energy's water use, such as incentives for adopting less water intense energy generation technologies. Another option is to make freshwater available for the energy sector (e.g., through allocations, permits, or facilitating water trading); however, the majority of water quantity allocation and permitting decisions are up to the states. An additional option is improving data and analysis on energy's water use to better inform decision-making (e.g., resource planning efforts and decision-support tools) and enhancing the availability and dissemination of water-efficient technological alternatives. These approaches are presented in Table 2, with examples of each from legislation reported by a congressional committee of the 111th Congress.

These options are not mutually exclusive and different options may be more or less appropriate and attractive for different components of the energy sector's water demand. These options also represent different potential roles and costs for federal and state governments, the energy sector, and energy consumers.

No entity has performed a comparative analysis of these policy options using multiple criteria (e.g., cost-effectiveness; who bears the cost; risks, reliability, and vulnerability; opportunity costs; role of state entities; role of federal entities). One of the challenges of a federal response to energy's water demand is that the concerns, policy options, and technological options vary greatly by region.[55]

Whether these policy alternatives should be pursued at the local, state, or national level depends in part on perspectives on the appropriate role of each level of government. Perspectives on the policy options related to energy's water demand also are influenced by long histories of regulation, management, promotion, and oversight of the nation's energy and water resources and infrastructure. Current and evolving conditions also play a role: providing greater access to water for the energy sector (which is primarily up to the states) may be difficult and controversial in water-scarce, drought-prone, or environmentally sensitive areas, especially with climate change anticipated to affect the availability and reliability of water in some regions.

Table 2. Sample Legislative Responses to Energy's Water Demand

Response	Examples in Committee-Reported Legislation in the 111th Congress
No Federal Action Directed at Managing Water Use	Existing federal energy (including conservation and efficiency policies), climate, agricultural, and environmental laws and policies indirectly, but at times significantly, shape the energy sector's water use.
Minimize Energy Sector's Growth in Water Use	
Promote water-efficient energy through standards, regulations, or incentives	Could be an outcome related to increased adoption of electric generation from wind, photovoltaics, or natural gas—which may occur as a result of energy markets, cap and trade legislation (e.g., H.R. 2454, American Clean Energy and Security Act of 2009, passed House and placed on Senate Legislative Calendar), or renewable electricity standards (e.g., S. 1462, American Clean Energy Leadership Act of 2009, reported by Senate ENR)— depending on how the energy industry responds to the legislation.
Promote use of technologies to reduce energy's water use though incentives or regulations	No explicit mention in legislation reported in the 111th Congress; however, state and federal energy permitting actions have required or promoted actions to reduce water use (e.g., roughly half of fast-tracked CSP development projects on federal lands are adopting dry cooling because of concerns about freshwater demands of other cooling options).
Promote conservation and efficiency measures through incentives or regulations	§ 2 of S. 3396, Supply Star Act of 2010 (reported by Senate ENR), would establish a federal program to promote resource-efficient industrial supply chains (such sums as necessary, $35 million total for five-year CBO authorization estimate). § 147 of Title I, Subtitle D, "Energy and Water Integration," in S. 1462 (reported by Senate ENR) would direct the Secretary of Energy to create a competitive energy-water conservation grant program for local governments, water agencies, and tribes ($100 million for each fiscal year from FY2010 to FY2015).
Develop and assess technologies to reduce energy's water use	§ 142 of S. 1462 would require a federal study to identify alternatives to optimize water and energy efficiency in the production of electricity (such sums as necessary). § 162 of S. 1733, Clean Energy Jobs and American Power Act (reported by Senate EPW), would require EPA to create a "1,000,000,000 Gallon Challenge Grant Program" to fund projects that have the potential for producing significant quantities of biofuels from renewable biomass, where surface and groundwater use and impacts are considered in determining whether the biofuel is renewable ($500 million for FY2010 through FY2014). § 196 of Title I of H.R. 2454 and § 152 of S. 1733 would authorize the Secretary of Energy to provide nonprofit entities with grants to conduct competitive programs to support start-up businesses in clean energy, including water conservation ($20 million).
Improve Energy Sector Access to Water	
Allocate sustainably available water	Largely state responsibility; no explicit mention in federal legislation, but some federal actions underway (e.g., review of requests for surplus water contracts under Flood Control Act of 1944, 33 U.S.C. § 708). Also, already authorized efforts, if completed, could inform allocation decisions; for example, § 9503 of Subtitle IV, "Secure Water," in P.L. 111-11 authorizes the assessment and development of strategies to address water quantity effects of global climate change in Bureau of Reclamation service areas.

Table 2. (Continued)

Response	Examples in Committee-Reported Legislation in the 111th Congress
Facilitate transfer of water saved in non-energy sector	§ 9504 of P.L. 111-11 establishes a program to provide grants or enter into cooperative agreements to (among other purposes) conserve water, increase water use efficiency, and facilitate water markets in the 17 Reclamation states ($200 million).
Inform Decisions and Support Research Expanding Water-Efficient Energy Options	
Data and Assessments	§ 9508 of P.L. 111-11 requires the Secretary of the Interior to establish a national water availability and use assessment program that reports on significant trends, including significant changes in water use due to the development of new energy supplies ($20.0 million for each of FY2009 through FY2013). § 3 of H.R. 3598, Energy and Water Integration Act (passed House), would create a council to recommend actions that promote improved energy and water resource data collection, and to conduct workshops to promote data exchange ($5 million for each of fiscal years 2011 through 2015).
	§ 149 of S. 1462 would mandate a study, led by the Department of Energy, examining industrial water use, peak energy use in water treatment and delivery, and energy "embedded" in water (no authorization of appropriations specified). § 9505 of P.L. 111-11, requires the Secretary of Energy to assess federal hydroelectric power production to water supply risks posed by global climate change (such sums as necessary).
Research	§ 2 of H.R. 3598 would require DOE to develop a strategic plan for energy-water research needs within DOE research activities ($60 million for each of FY2011 through FY2015). § 171 of Subtitle H, Title I, of H.R. 2454 would direct the Secretary of Energy to establish energy innovation hubs to conduct research, including research that enhances water security (no specific sum set).
Integrated energy - water planning	No explicit mention in legislation in the 111th Congress
Technical assistance	§ 148 of S. 1462 would require the Secretary of Energy to create a technical assistance program for reducing energy use by rural water utilities ($7 million for each of FY2010 through FY2015). § 10 of H.R. 1145, National Water Research and Development Initiative Act of 2009, would require the EPA Administrator to establish a pilot program for energy audits of water related infrastructure (no amount specified for this section, entire bill has $2 million for each of FY2010 through Fy2014).

Source: CRS.

Notes: ENR = Energy and Natural Resources Committee, EPW = Environment and Public Works Committee, EPA = U.S. Environmental Protection Agency.

States have traditionally had primacy in water allocation, so decisions about permitting energy's water use largely have been non-federal. State and federal laws and policies can affect the ease or difficulty of, and incentives for, transferring water from existing uses to energy.

The federal government can, if it chooses, promote change in state water laws, institutions, and decisionmaking.[56] Some view this as infringing on states' rights.[57] Similarly, while private entities make many of the decisions on the energy sector's water use, the public sector influences these decisions through numerous routes (e.g., tax incentives, loan guarantees, permits, regulations, planning, education); this influence can come from local, state, or federal policies. The federal role in water resource allocation and management increases as the federal interest increases—for example, when the use occurs on federal lands. Competing water demands, including those from the energy sector, are raising questions for federal agencies about the operations of federal facilities. For instance, can water delivered to Bureau of Reclamation contractors (e.g., water delivered to irrigation districts in California's Central Valley) be used not for agriculture, but for energy development (e.g., evaporative cooling of a concentrating solar power facility)? Can water from an Army Corps of Engineers dam that has multiple purposes (e.g., flood control, navigation, and/or municipal or agricultural water supply) be used for the oil and gas development, and if so, how much water can the Corps provide under existing authorities?

Pursuit of Water Stored Behind Federal Dams for Oil and Gas Water Supply

Western North Dakota illustrates some stakeholders' interest in using federal infrastructure to satisfy the energy sector's water demand. The region has experienced substantial expansion in oil and gas production since 2008. The industry's largely new water demand is anticipated to soon reach 24 million gallons per day (27,000 acre-feet annually). At that point, it would represent one-third of total water usage in the 11 counties in Western North Dakota that are producing oil and gas and located in proximity to a lake formed by a federal dam on the Missouri River.

In December 2010, the Corps released a draft report, *Garrison Dam/Lake Sakakawea Project, North Dakota, Draft Surplus Water Report* (Dec. 2010), that if finalized would allow the Secretary of the Army to meet the industry's water demand for up to 10 years by allowing withdrawals from Lake Sakakawea at Garrison Dam (ND). While other options for accessing the water sought by the oil and gas industry are available, the draft report finds that the Corps' provision of water from the lake would be the least costly means of addressing the water demands of the oil and gas industry and other water users in the region.

OBSERVATIONS AND CONCLUDING REMARKS

The energy sector has long been a major water user, so why the current concern? Major energy trends are pushing the energy sector to become more dependent on, and therefore vulnerable to, freshwater availability. This is occurring at a time of increasing concerns about the adequacy and reliability of freshwater supplies due to population growth and climate change. Energy resource and technology paths chosen and capital investments made in the near term are likely to establish long-term trajectories for energy's water use.

Many of the current trends are in part driven by federal policy. The federal government partially shapes the energy sector and at times defines a vision for the nation's energy future (e.g., biofuel production targets). Some stakeholders have raised concerns about the feasibility and consequences of meeting various energy targets and policy proposals, including concerns about physical inputs like water, land, materials, and rare minerals.[58]

Because affordable freshwater is a finite resource, commitments of water supplies for the energy sector reduce availability for other sectors and for ecosystems. Local or regional competition for water is often what makes energy's water demand significant; at the same time, it is the regional and local scale of water resources and how they are managed that often complicates many federal water-related actions.

The federal role in energy supply and demand raises questions about the policy direction for meeting and managing energy's water needs, among them: If energy security is a national security issue, is energy's water use by association a national security issue? Would this be a justification for federal spending on energy and water efficiency measures?

Water supply concerns are not only being raised in the context of energy. There also is growing concern about water availability and aquatic conditions for meeting the demands of the agricultural and municipal sectors and the needs of ecosystems and threatened and endangered species; these concerns are raised particularly in the context of droughts and impacts of climate change on water resources.

Given available freshwater resources, a challenge for the nation will be to cost-effectively, sustainably, and reliably meet energy demands while satisfying agricultural, municipal, and industrial water demand, as well as ecosystem needs.

This challenge raises fundamental and controversial questions about how U.S. freshwater resources are allocated and used for various purposes, and about the availability and value of water in different sectors of the economy and in the environment. At issue for Congress is the role that the federal government and federal funding plays in shaping, meeting, and managing energy's water demand, while accounting for the significant role of the states and private sector in water use decisions.

APPENDIX A. ENERGY'S WATER CONSUMPTION TRENDS

Ideally, policy and decision makers should know how energy choices and policies compare across a wide array of parameters (e.g., cost, reliability, dispatchability, emissions, land requirements, water use) under different scenarios.

That is, informed decisions would require water use data, analyses of least-cost water and energy conservation and efficiency actions, understanding of other water uses and their costs and benefits, life-cycle assessments of energy's water uses and risks, and more. Such assessments are not available. While not addressing this shortfall, Table A-1 and Table A-2 summarize the freshwater impacts of numerous shifts in transportation fuels and the electricity sector, respectively.

Without such multi-variable assessments, it is difficult to understand the full implications of energy decisions. While this report focuses on energy's water use, there are a host of other factors to consider when analyzing energy policy tradeoffs.

For example, fuel and technologies for generating electricity are not equally dispatchable; that is, is they are not equally able to increase or decrease generation, or to be brought online or shut down to match demand. Thermoelectric facilities using fossil fuel or geothermal sources are advantageous because they can be operated to produce electricity constantly or dialed up or down with demand.

While electricity from hydropower, tidal, and wave energy generally can be produced fairly predictably, often the timing of operations is subject to the intensity of tides and waves, to water conditions such as drought or large storms, and to environmental restrictions regarding potential impacts on marine life and ecosystems.

In contrast, until more advanced storage technologies become commercially competitive, photovoltaics and wind, which require very little water, remain intermittent sources that generally cannot be dispatched when the sun is not shining or the wind is not blowing.

Table A-1. Domestic Freshwater Impacts from Transportation Fuel Shifts

Shifts in Imported Fuel Use	Change in Domestic Freshwater Consumption	Impairments of domestic water quality and/or aquatic ecosystems
Domestic fuel sources displace foreign fuels	Little change or ↑ depending on domestic fuel	↑
Domestic oil using enhanced oil recovery displaces foreign gasoline	↑	↑ from produced water
Domestic offshore oil displaces foreign oil	Little change	↑ from spill risk
Fuel conservation (e.g., resulting from increased fuel efficiency standards) decreases oil imports	Little change	Little change
Imported biofuels displace foreign gasoline	Little change	Little change

Source: CRS.

Table A-2. Domestic Freshwater Impacts from Electricity Sector Shifts

Shifts in Electricity Generation	Change in U.S. Freshwater Consumption	Impairment of Domestic Water Quality and/or Aquatic Ecosystems
Fuel Source Shift		
Nuclear displaces coal	Unclear with current data	Shift in pollutants generated during fuel extraction, processing, and electricity generation
Concentrating solar power (with freshwater cooling) displaces coal	↑	↓
Domestic natural gas displaces coal	↓	Shift from nonpoint source to more point source pollution
Hydropower without new reservoirs displaces coal	↓	Shift from nonpoint source pollution to potential harm to aquatic ecosystem and species
Photovoltaics displace coal	↓	↓
Wind displaces coal	↓	↓
Ocean/tidal/hydrokinetic displace coal	↓	Potential for harm to aquatic ecosystem and species
Geothermal displaces coal	↓	↓
Electricity Sector Shifts		
CCS added to a power plant	↑	unknown

Shifts in Electricity Generation	Change in U.S. Freshwater Consumption	Impairment of Domestic Water Quality and/or Aquatic Ecosystems
Dry cooling replaces evaporative cooling	⇩	⇩
Electricity conservation decreases electricity consumption	⇩	⇩

Source: CRS.

APPENDIX B. WATER USE BY TRANSPORTATION FUELS

D. Elcock's study projects that the energy sector's consumptive water use will increase from 6 billion gallons per day (bgd) in 2005 to 10 bgd in 2030 in the areas of fossil fuel mining, production, and processing and bioenergy.[59] (See Figure B-1.) The growth is dominated by water consumed for bioenergy.[60] The Elcock study and other sources show that the effect of bioenergy on energy's water use was already felt between 2005 and 2010.[61]

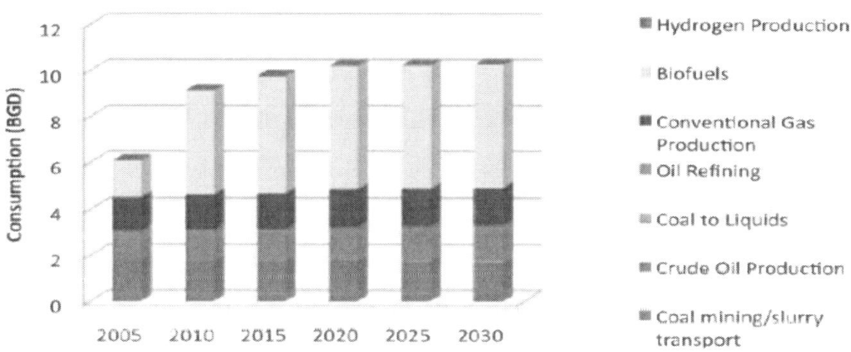

Source: Elcock 2010.

Figure B-1. Biofuels Dominate Energy Production's Projected Water Consumption.

The projected water demand by bioenergy could drop, potentially significantly, if less-water intense bioenergy is developed and adopted. This projection covers most of the water used to fuel transportation and the water used for obtaining and preparing fossil fuels for use in either transportation or the electricity sector. Figure B-2 provides data from research on the comparative water intensity of different U.S. transportation fuels. Note that the figure has a logarithmic scale, and that the water consumed traveling on irrigated biofuels is orders of magnitude larger than the water embedded in other fuels.

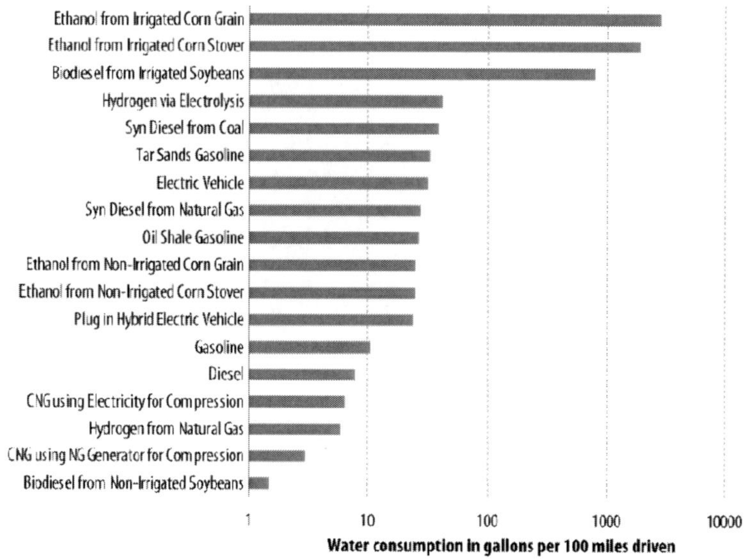

Source: CRS, modified from M. Mantell, "Abundant, Affordable, and Surprisingly Water Efficient," presented at 2009 GWCP Water/Energy Sustainability Symposium in Salt Lake City, September 13-16, 2009. Original data from C. King and M. Webber, "Water Intensity of Transportation," *Environmental Science & Technology*, vol. 42, no. 21 (2009), available at http://pubs.acs.org/doi/pdf /10.1021/es800367m.

Notes: NG = natural gas; CNG = compressed natural gas. Note the logarithmic scale.

Figure B-2.Water Intensity of Transportation Fuels.

APPENDIX C. WATER USE FOR ELECTRICITY GENERATION

Trends in the electricity sector contributing to increased water consumption currently overwhelm actions that reduce water consumption. As a result, local water challenges presented by electric generation are becoming more common in many water-constrained areas. These challenges may drive action to reduce electricity's freshwater footprint, such as the adoption of water-efficient electricity generation and water-efficient thermoelectric cooling options, or the use of impaired water. This appendix describes the available data on the electricity sector's water consumption and data on the different electricity generation's water use. Then it presents alternative thermoelectric cooling choices that may reduce electricity's water use. Lastly, it discusses the tradeoffs between water use and other characteristics (e.g., environmental impacts, reliability, dispatchability) of electric generation from hydropower facilities, photovoltaic solar and wind installations, and geothermal resources.

Data on Electricity's Water Use

Available estimates for the electricity sector's water consumption focus primarily on thermoelectric cooling water needs because cooling dominates water use during generation. Other aspects of electric production (e.g., fuel mining and processing) also use water. (See Table C-1.) Data and projections on the water consumed in the mining, production, and processing of fuels for generating electricity are bundled with the production of fossil fuels used in the transportation sector, like the Elcock study. The projections in Figure 2 for thermoelectric cooling come from the Elcock study, which was based on projections in a 2007 report by NETL.[62] The NETL projections are limited to coal, natural gas, and nuclear-powered facilities; that is, they do not calculate water use changes that would occur from shifts in the electricity sector broadly (e.g., adoption of more wind or photovoltaics) or shifts in thermoelectric generation more specifically (e.g., concentrating solar power or biomass generation). Elcock assumed that biomass generation for electricity production would be based in regions not requiring irrigation.[63]

No analyses of electricity's total water consumption or how it may change under different policies exist; this type of analysis is difficult to perform without reliable data on water intensities of different electricity generation technologies and fuels. No authoritative comparison of the various water intensities of electricity generation options exists. Figure C-1 and Figure C-2 illustrate the best available data for average water intensities for various electricity alternatives, with and without carbon mitigation using CCS, respectively. These figures are based on Table C-1, which represents the best available data on water consumed in producing electricity using various fuels and technologies.[64] The data in Figure C-1 are imperfect and come from multiple sources, raising questions about whether the data can be used to make accurate comparisons across technologies.

Table C-1. Water Consumption for Electricity Generation by Fuel Source

Generation Technology and Fuel	Water for Fuel Mining, Production, or Processing (gal/MWh)	Evaporative Cooling Water at Power Plant (gal/MWh)	Other Water Used for Power Plant Operations (gal/MWh)	Avg. Total Water Intensity (gal/MWh)	Data Sources
Biomass/Waste					
Non-irrigated biomass	0	300-480	30	420	DOE 2006
Non-irrigated biomass with CCS	0	Not available	30	—	DOE 2006
Irrigated biomass	Highly variable	300-480	30	—	DOE 2006
Municipal waste	Not available	300-480	30	420	DOE 2006
Coal					
Conventional/Subcritical coal	5-74	449	68	557	DOE 2006; NETL 2009
Conventional/Subcritical coal with CCS	5-74	884	101	—	DOE 2006; NETL 2009
Supercritical coal	5-74	392	59	491	DOE 2006; NETL 2009
Supercritical coal with CCS	5-74	759	86	—	DOE 2006; NETL 2009
Ultra-supercritical Coal Ultra clean coal	Not available	Not available	Not available		
	Not available	Not available	Not available		
Coal IGCC (slurry fed)	30-70	290	19	359	DOE 2006; NETL 2009
Coal IGCC (slurry fed) with CCS	30-70	355	97	502	DOE 2006; NETL 2009
Coal IGCC (dry fed)	5-74	243	53	336	DOE 2006; NETL 2009
Coal IGCC (dry fed) with CCS	5-74	355	120	515	DOE 2006; NETL 2009
Geothermal					
Enhanced geothermal	0	Dry cooling	290-720	585	Argonne 2010
Geothermal binary	0	Dry cooling	80-270	175	Argonne 2010
Geothermal flash	0	Dry cooling	5-10	8	Argonne 2010

Generation Technology and Fuel	Water for Fuel Mining, Production, or Processing (gal/MWh)	Evaporative Cooling Water at Power Plant (gal/MWh)	Other Water Used for Power Plant Operations (gal/MWh)	Avg. Total Water Intensity (gal/MWh)	Data Sources
Hydroelectric			Evaporation varies with site		NREL 2003
Natural Gas					
Natural gas combined-cycle Natural gas combined-cycle with CCS	11	192 338	0 0	203	DOE 2006; NETL 2009
	11			349	DOE 2006; NETL 2009
Nuclear					
Nuclear steam cycle	45-150	720 (cooling pond)	30	848	DOE 2006
Nuclear boiling water	45-150	Not available	—	—	DOE 2006
High temperature gas cooled nuclear	45-15	Not Available	—	—	DOE 2006
Small modular nuclear	Not Available	Dry cooling possible	Not available	—	
Ocean/Tidal	0	Not applicable	Not available	0	
Oil					
Oil	No data	300-480	Not available	—	DOE 2006
Oil from enhanced oil recovery	Highly variable, potentially significant	300-480	Not available	—	DOE 2006
Solar					
CSP – solar trough	0	760-920	80	920	DOE 2006; DOE n.d.
CSP – solar tower	0	750	90	840	DOE 2006; DOE n.d.
Dish Engine Solar	0	Air cooled	4	4	NREL 2002

Table C-1. (Continued)

Generation Technology and Fuel	Water for Fuel Mining, Production, or Processing (gal/MWh)	Evaporative Cooling Water at Power Plant (gal/MWh)	Other Water Used for Power Plant Operations (gal/MWh)	Avg. Total Water Intensity (gal/MWh)	Data Sources
Photovoltaics	0	Not applicable	4, higher for locations requiring more panel washing	4	NREL 2002
Wind – onshore or offshore	negligible		0		DOE 2008

Sources: Argonne National Laboratory released for comment the following report: ANL, *DRAFT Water Use in the Development and Operation of Geothermal Power Plants*, Sep. 2010. *Energy Demands on Water Resources: Report to Congress on the Interdependency of Energy and Water*, Dec. 2006. DOE, *Concentrating Solar Power Commercial Application Study: Reducing Water Consumption of Concentrating Solar Power Electricity Generation*, Washington, DC, no date (n.d.). NETL, *Estimating Freshwater Needs to Meet Future Thermoelectric Generation Requirements (2009 Update)*, 2009. NREL, *Fuel from the Sky: Solar Power's Potential for Western Energy Supply*, NREL/SR-550-32160 July 2002. P. Torcellini et al., *Consumptive Water Use for U.S. Power Production* (NREL), 2003.

Notes: Evaporative cooling for thermoelectric facilities is assumed, unless otherwise noted in the table. Data do not include water consumed in manufacturing or construction. While the total column represents the average total water consumption for a unit of electricity, the water for fuel production may occur at a different location than the water consumed at the power plant.

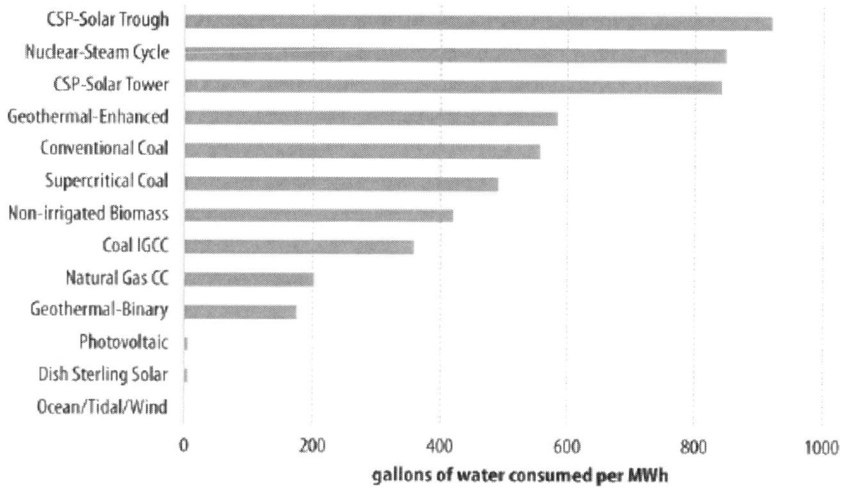

Source: CRS. Data represent evaporative-cooled thermoelectric, except for geothermal, see Table C-1. IGCC = Integrated Gasification Combined-Cycle; CC = Combined Cycle.

Figure C-1.Water Intensity of Electricity Generation Alternatives without CCS.

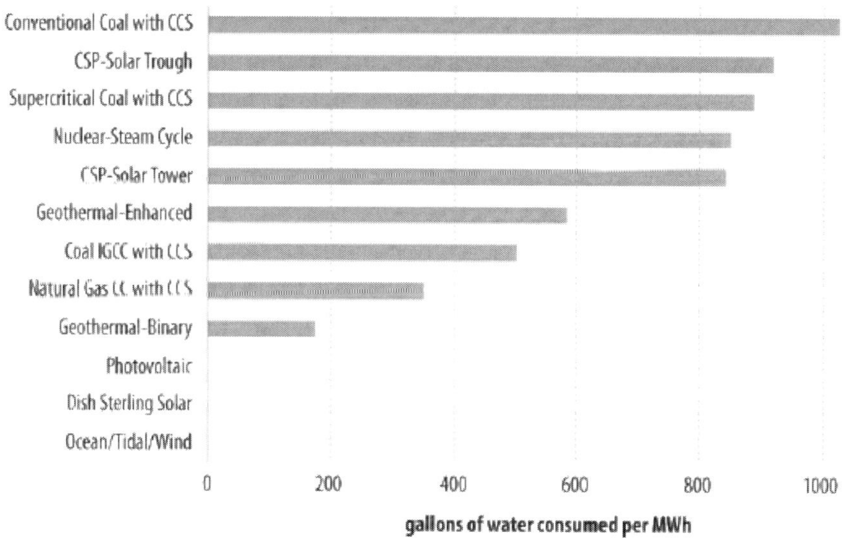

Source: CRS. Data represent evaporative-cooled thermoelectric, except for geothermal, see Table C-1.

Figure C-2.Water Intensity of Electricity Generation Alternatives with CCS.

Thermoelectric Cooling: Emerging Alternatives

The withdrawal and water quality impacts of once-through cooling have resulted in newer power plants generally using evaporative cooling.[65] Emerging cooling technologies have the potential to use much less freshwater than once-through or evaporative cooling. These include dry cooling (previously discussed), hybrid dry-wet cooling, cooling with fluids other than freshwater, and more innovative technologies (e.g., wet-surface air coolers, advanced wet cooling). However, these alternatives have their own costs and disadvantages.

A DOE report found that dry cooling could reduce water consumption to roughly 80 gal/MWh for solar troughs and 90 gal/MWh for solar towers.[66] While they consume less water, dry and hybrid cooling systems have financial as well as efficiency costs. Total annualized costs for dry cooling tower systems can be four times those of evaporative cooling tower systems.[67] Due to the higher cooling and lower generation efficiency costs, the cost of electricity from a dry cooled plant may be 10% higher than a similar wet cooled plant.[68] Dry cooling uses fans to blow air for steam condensation. While power plants with dry cooling use considerably less water, dry cooling is less effective at cooling the power plant than evaporative cooling, thus reducing electric generation at the facility. The DOE report also found that electricity generation at a dry-cooled facility dropped off at ambient temperatures above 100°F. For a solar parabolic trough facility in the Southwest, the benefit in the reduction in water consumption from dry cooling resulted in cost increases of 2% to 9% and a reduction in energy generation of 4.5% to 5%.[69] The cost and energy generation penalties for dry cooling depend largely on how much time a facility has ambient temperature above 100°F. Dry cooling would reduce generation on the same hot days when summer peak electricity demand is greatest.

Hybrid wet-dry cooling attempts to balance water consumption with power generation efficiency; it remains under development for commercial scale applications. To weigh the tradeoffs in energy generation, cost, and water use, DOE researched hybrid cooling processes that combine dry and evaporative cooling. The hybrid system consists of parallel evaporative and dry cooling facilities, with the evaporative cooling operating only on hot days. By using dry cooling generally and evaporative cooling above certain ambient temperatures, losses of thermal efficiency from dry cooling can be reduced. How often the evaporative cooling is used determines how much water is consumed and the effect of hot days on thermal efficiency. DOE found that a hybrid cooling system in the Southwest using 50% of the water of evaporative cooling would maintain 99% of the performance of an evaporative-cooled facility. A hybrid cooling system using 10% of the water of evaporative cooling would maintain 97% of the energy performance.

Another means for decreasing freshwater impacts is employing alternative water sources for evaporative-cooling.[70] These alternative water sources include effluent from wastewater treatment plants[71] or other reclaimed or impaired water, such as brackish or low-quality groundwater[72] and mine pool water.[73] These alternative water sources, however, may lead to additional scaling, corrosion, and fouling of cooling equipment or require pretreatment before cooling use.

Additional research may be able to improve the viability of saline water cooling. Availability of alternative water sources in proximity to electricity generation facilities is a potential limitation to their use. Understanding of the opportunities for brackish cooling alternatives is likely to be improved when the assessment of brackish groundwater authorized by Section 9507 of P.L. 111- 11, the Omnibus Public Land Management Act of 2009, becomes available.

Tradeoffs of Select Electricity Generation Technologies

Hydroelectric Generation

Hydroelectric power is produced when water passes through a turbine. Turbines for large-scale hydroelectric generation are located at dams. Electricity at U.S. hydropower facilities is produced with relatively low greenhouse gas emissions. However, hydropower's environmental effects can be significant. Conventional hydropower development through dam building often significantly alters river ecosystems, harming many indigenous species. Drought and changes to hydrology, such as possible reduction in snowpack under a changing climate, can reduce electricity generation at hydropower facilities because of the effects on reservoir operations and levels.

Constructing new large dams is contentious; therefore, efforts to identify opportunities for increasing hydropower generation have focused on smaller-scale opportunities or improved efficiency and expansion of hydropower at existing facilities.[74] The Electric Power Research Institute estimates a potential capacity hydropower gain of 10 GW by 2025 as feasible, without the construction of new large dams.[75] Six western states—Alaska, Washington, Oregon, California, Idaho, and Montana—have the highest potentials.[76]

Despite this identified potential, little additional hydropower generation has been installed in recent years for a number of reasons (e.g., hydropower and water-related permit and regulatory requirements, and public concerns and perceptions about environmental effects). Similarly, the number of applications for FERC preliminary permits for pumped storage has increased substantially in recent years; however, the proposals have not proceeded to construction.

Photovoltaic Solar and Wind

Renewable electricity technologies that do not use thermoelectric processes may have minimal water requirements for electricity generation. Wind turbines and solar photovoltaic (PV) panels, for example, require small volumes of water for cleaning, but otherwise use no water. However, the minimal water intensity of wind and PV comes with tradeoffs. Transmission constraints;[77] cost; and regulatory, technical, and operational factors currently restrict the extent to which solar and wind resources can be exploited to meet electricity demand. As previously noted, wind and PV are intermittent electricity sources. Some storage options for these intermittent technologies exist (e.g., wind used in conjunction with a pumped storage hydropower facility); however, their applications are limited and intermittency continues to limit generation from wind and PV.[78] Electricity from PV is also currently more expensive than electricity from CSP, although electricity from wind is less expensive than electricity from CSP.

Geothermal

There are several ways to use geothermal energy: electricity generation (discussed herein) as well as direct-use (recovering water heated by the earth) and heat pumps (using the earth's heat to cool/heat buildings). Traditional geothermal power production uses naturally occurring convective hydrothermal sources in hot rock formations to produce steam to run a thermoelectric power plant's turbines (i.e., a geothermal flash system). Alternatively, for lower temperature geothermal resources, a second working fluid is heated by the geothermal water using a heat exchanger; it is the working fluid that drives the turbines (i.e., a geothermal binary system). Finally, because the majority of hot rock is dry, electricity can also be generated by injecting water into fractured rock to be heated (i.e., an enhanced geothermal system).[79]

The water is then injected back into the rock formation to create a relatively closed-loop system. Because the geothermal or injected water is an essential component of geothermal electricity generation and the size of the plants are generally smaller than 50 MW, dry cooling is becoming the standard for new geothermal facilities. Smaller power plants are generally easier to dry-cool than larger plants.

A 2008 USGS study estimated that the United States has geothermal resources sufficient to supply half of the nation's electric generation needs, assuming enhanced geothermal systems could be successfully developed and deployed at a commercial scale.[80]

That is, much of the geothermal potential shown on maps of geothermal resources[81] requires water to be injected (and therefore consumed) to exploit the geothermal energy. Enhanced geothermal power plants require relatively little land and can be used in coproduction with enhanced oil recovery to lengthen the lifespan of oil fields.[82] These enhanced systems are an emerging technology, so more research and development are needed for large-scale commercial deployment.

Current research is investigating the possibility of replacing water with carbon dioxide as the working fluid, which would significantly reduce water usage and would be a means of carbon sequestration.

End Notes

[1] Thermoelectric power plants burn or react fuel to generate steam, which turns a turbine connected to a generator that produces electricity. Cooling water is used to condense the steam into boiler feed water, so the process can be repeated. See Appendix C for more information on thermoelectric cooling technologies and water use.

[2] U.S. Geological Survey, *Estimated Use of Water in the United States in 2005* (Circular 1344: 2009).

[3] D. Elcock, "Future U.S. Water Consumption: The Role of Energy Production," *Journal of the American Water Resources Association*, vol. 46, no. 3 (June 2010), pp. 447-480, hereafter referred to as Elcock 2010.

[4] See CRS Report RS21421, *Mountaintop Mining: Background on Current Controversies*, by Claudia Copeland

[5] See CRS Report R41311, *The Deepwater Horizon Oil Spill: Coastal Wetland and Wildlife Impacts and Response*, by M. Lynne Corn and Claudia Copeland.

[6] DOE, National Energy Technology Laboratory (NETL), *Estimating Freshwater Needs to Meet Future Thermoelectric Generation Requirements (2009 Update)*, p. 64, http://www.netl.doe.gov/energy-analyses/pubs/ 2009%20Water%20Needs%20Analysis%20-%20Final%20 (9-30-2009).pdf, hereafter referred to as NETL 2009.

[7] EIA, *Annual Energy Outlook 2010 Early Release*, Washington, DC, http://www.eia.doe.gov /oiaf/aeo/aeoref_tab.html.

[8] For more on oil shale and water rights, see CRS Report RS22986, *Water Rights Related to Oil Shale Development in the Upper Colorado River Basin*, by Cynthia Brougher.

[9] U.S. Department of Agriculture (USDA) Forest Service, *2000 RPA Assessment of Forest and Range Lands*, FS-687, February 2001, p. 14.

[10] See U.S. Climate Change Science Program and the Subcommittee on Global Change Research, *Effects of Climate Change on Energy Production and Use in the United States*, Synthesis and Assessment Product 4.5, February 2008, pp. 31-32, hereafter referred to as U.S. Climate Science Program 2008.

[11] M. Hightower and S.A. Pierce, "The Energy Challenge," *Nature*, vol. 452, no. 7185 (March 20, 2008), pp. 285-286.

[12] Tennessee Valley Authority, "President's Report" presentation, Nov. 2010, p. 88, http://www.tva.gov/abouttva/ board/pdf/11-4-2010_board_final.pdf.

[13] Ibid, p. 89.

[14] U.S. Climate Science Program 2008, p. 45, found that "hydroelectric power generation can be expected to be directly and significantly affected by climate change...with potential for production decreases in key areas such as the Columbia River Basin and Northern California." Hydroelectric generation also may decrease if dams are reoperated to maximize stored water and minimize losses to evaporation, in areas where conserving available freshwater supplies may become more critical.

[15] M. Furniss et al., *Water, Climate Change and Forests: Watershed Stewardship for a Changing Climate*, U.S. Department of Agriculture, Forest Service, Pacific Northwest Research Station, PNW-GTR-812, Portland, OR, June 2010, http://www.fs.fed.us/pnw/pubs /pnw_gtr812.pdf, hereafter referred to as Forest Service 2010.

[16] U.S. Climate Science Program 2008, p. 38.

[17] For a discussion of data gaps related to thermoelectric power plants, see U.S. Government Accountability Office (GAO), *Energy-Water Nexus: Improvements to Federal Water Use Data Would Increase Understanding of Trends in Power Plant Water Use*, GAO-10-23, Oct.16, 2009.

[18] Past forecasts of national water use have proven highly inaccurate. The draft report cited below includes a graph showing a wide variation of previously projected freshwater withdrawals. It further illustrates that all but one overestimated actual water withdrawals: U.S. Army Corps of Engineers, Institute of Water Resources, *Plan of Study: National Water Demand and Availability Assessment*, DRAFT, Alexandria, VA, Feb. 1995, p. 7. Some recent reports, like Forest Service 2010, indicate a role for scenario-based forecasts.

[19] W. B. Solley et al., *Estimated Use of Water in the United States in 1995* (USGS Circular 1200, 1998), http://water.usgs.gov/watuse/pdf1995/html/. Hereafter referred to as USGS 1998.

[20] The Forest Service 2010 water assessment is anticipated to assess the vulnerability of the coterminus United States to water supply shortages over the next 50 years, including alternative scenarios for climate change. It is not breaking out energy's water use or evaluating the sensitivity of the assessments to various policy and technology choices. For more information on this assessment, see http://www.fs.fed.us/rm/human /2010_rpa_ assessment.shtml.

[21] Elcock 2010; D. Elcock, *Baseline and Projected Water Demand Data for Energy and Competing Water Use Sectors*, Argonne National Laboratory, Environmental Science Division, November 2008.

[22] U.S. Dept of the Interior, "Secretary Salazar Launches New Regional Climate Science Center and Water Census at Meeting of Colorado River Basin Water Leaders," press release, Oct. 20, 2010, http://www.doi.gov/news/pressreleases/ Secretary-Salazar-Launches-New-Regional-Climate-Science-Center-and-Water-Census-at-Meeting-of-Colorado-RiverBasin-Water-Leaders.cfm.

[23] For more information see, the Western Governors' Association, *Regional Transmission Expansion Project*, http://www.westgov.org/index.php?option=com_content&view=article&id=311&Itemid=81.

[24] USDA Forest Service, *Past and Future Freshwater Use in the United States: A Technical Document Supporting the 2000 USDA Forest Service RPA Assessment*, RMRS-GTR-39, Sept. 1999, p. 44.

[25] The extent to which increased water use for bioenergy results in a net increase in national water consumption remains the subject of debate.

[26] Calculated using data tables for Yi-Wen Chiu, Brian Walseth, and Sangwon Suh, "Water Embodied in Bioethanol in the United States," *Environmental Science & Technology,* vol. 43, no. 8 (2009), pp. 2688-2692, hereafter referred to as Chiu 2009; and Kansas Department of Agriculture, Division of Water Resources, *Water Use in Kansas*, http://www.ksda.gov /appropriation/content/116. Insufficient data are available to analyze more local effects on water use (e.g., county level water use). Chiu 2010 includes water used for irrigation and for converting the feedstock into fuel; it does not include precipitation, soil moisture, and other water indirectly consumed by the feedstock.

[27] Forest Service 2010, pp. 15-16, and 24.

[28] Ibid.

[29] DOE, *20% Wind Energy by 2030: Increasing Wind Energy's Contribution to U.S. Electricity Supply*, July 2008, p. 185, http://www1.eere.energy.gov/windandhydro/pdfs/41869.pdf, hereafter referred to as DOE 2008. In the report, the Midwest/Great Plains region included Illinois, Indiana, Iowa, Kansas, Michigan, Minnesota, Missouri, Nebraska, North Dakota, Ohio, Oklahoma, South Dakota, Texas, and Wisconsin.

[30] EIA, Table 1.1, Net Generation by Energy Source, http://www.eia.gov/cneaf/electricity /epm/table1_1.html.

[31] For more on carbon capture, see CRS Report R41325, *Carbon Capture: A Technology Assessment*, by Peter Folger

[32] NETL 2009, p. 64.

[33] USGS 1998, p. 54.

[34] NREL 2003.

[35] The NREL 2003 report included data on the average water consumption per unit of hydroelectric generation by state; the water consumption intensity for the hydroelectric produced at these facilities, according to the study, can vary from a low of 2,000 gallons/megawatt-hour (MWh)(Nebraska) to a high of 154,000 gallons/MWh (Kentucky). The report calculated the weighted average rate of water consumption per MWh at 18,000 gallons, with the variation in the rates across states being significant; the report calculated the national weighted average for thermoelectric generation at 470 gal/MWh.

[36] DOE, Office of Fossil Energy, *Modern Shale Gas Development in the United States: A Primer*, April 2009, p. ES-2, http://www.netl.doe.gov/technologies/oil /Shale_ Gas_Primer_ 2009.pdf.

[37] Ibid, p. 64. An Army Corps of Engineers draft report put the lower end of water use at 0.8 million gallons, with the lower end dominated by the shorter vertical wells. U.S. Army Corps of Engineers, Omaha District, *Garrison/Lake Sakakawea Project, North Dakota, Draft Surplus Water Report*, Omaha, NE, Dec. 2010, p 3-2.

[38] M. Mantell, "Abundant, Affordable, and Surprisingly Water Efficient," presented at 2009 GWCP Water/Energy Sustainability Symposium in Salt Lake City, September 13-16, 2009, available at http://www.energyindepth.org/wpcontent/uploads/2009/03 /MMantell_GWPC_

Water_Energy_Paper_Final.pdf; and EIA, *Average Heat Rates by Prime Mover and Energy Source*, January 2010, http://www.eia.doe.gov/cneaf/electricity/epa/epat5p4.html.

[39] DOE, *Energy Demands on Water Resources: Report to Congress on the Interdependency of Energy and Water*, Dec. 2006, http://www.sandia.gov/energy Congress-EWwEIAcomments-FINAL.pdf, hereafter referred to as DOE 2006.

[40] L. P. Galusky, *An Update and Prognosis on the Use of Fresh Water Resources in the Development of Fort Worth Basin Barnett Shale Natural Gas Reserves*, Texerra Engineering, prepared for Barnett Shale Education Council and Barnett Shale Water Conservation and Management Committee, Midland, TX, Nov. 4, 2009.

[41] Ibid.

[42] C.J. Vavra, "Desalination of Oil Brine," presentation, http://www.pe.tamu.edu/gpri-new/home/BrineDesal/ MembraneWkshpAug06/Burnett8-06.pdf

[43] CRS Report R40631, *Water Issues of Concentrating Solar Power (CSP) Electricity in the U.S. Southwest*, by Nicole T. Carter and Richard J. Campbell.

[44] Ibid.

[45] Dish engine CSP facilities employ engines instead of steam turbines. Mirrors concentrate sunlight to produce heat in a gas chamber connected to a piston and drive shaft. The drive shaft powers an electricity generator. Because of the high operating temperature and high efficiency of the motor, air cooling can be used with little compromise of overall electricity generation efficiency. This significantly reduces the water used to generate electricity compared to other CSP technologies. One dish engine facility with a capacity of 2 MW was put into operation in 2010 in Arizona; dish engine facilities of 750 MW and 850 MW are under development in southern California.

[46] This section on biofuels was written by Megan Stubbs, Analyst in Agricultural Conservation and Natural Resources Policy, Resources, Science and Industry Division (7-8707).

[47] For more information and a map of the Ogallala aquifer, see V. L. McGuire, *Changes in Water Levels and Storage in the High Plains Aquifer, Predevelopment to 2005*, USGS, USGS Face Sheet 2007-3029, 2007, http://pubs.usgs.gov/fs/ 2007/3029/.

[48] Calculated using supplementary data tables from Chiu 2009, and City of Fort Collins, *Water Supply and Demand*, Fort Collins, CO, http://www.fcgov.com/water/supply.php.

[49] S. E. Powers, R. Domingues-Faus, and P. J. J. Alvarez, "The Water Footprint of Biofuel Production in the USA," *Biofuels*, vol. 1, no. 2 (2010), pp. 255-260.

[50] GAO, *Energy Water Nexus: Many Uncertainties Remain about National and Regional Effects of Increased Biofuel Production on Water Resources* (GAO 10-116, Nov. 2009), p. 10. Agricultural runoff with excess nutrients from fertilizers used in the Mississippi River basin contributes significantly to hypoxic "dead zones" in the Gulf of Mexico. For more information on dead zones, see CRS Report 98-869, *Marine Dead Zones: Understanding the Problem*, by Eugene H. Buck.

[51] C.W. King,, M. E. Webber, and I.J. Duncan, "The Water Needs for LDV Transportation in the United States," *Energy Policy*, vol. 38 (2010), p. 1161, hereafter referred to as King 2010.

[52] In particular, the renewable fuels standard (RFS), a major federal incentive established in the Energy Independence and Security Act of 2007 (EISA, P.L. 110-140), establishes a goal of 36 billion gallons of biofuel production by 2022, including 16 billion gallons of cellulosic biofuels. For more information on the RFS, see CRS Report R40155, *Renewable Fuel Standard (RFS): Overview and Issues*. Also, the Food, Conservation, and Energy Act of 2008 (the 2008 farm bill, P.L. 110-246) supports the growth of the cellulosic industry through research programs, grants, loans, and tax credits. For more information on energy

provisions in the 2008 farm bill, see CRS Report RL34130, *Renewable Energy Programs in the 2008 Farm Bill*, by Megan Stubbs.

[53] K.R. Fingerman et al., "Accounting for the Water Impacts of Ethanol Production," *Environmental Resources. Letters*, vol. 5 (2010), http://iopscience.iop.org/1748-9326/5/1/014020/pdf/1748-9326_5_1_014020.pdf. Others also have raised concerns that large-scale deployment of open pond algae systems in arid areas may have a freshwater footprint if impaired waters are not used (C. Harto, R. Meyers, and E. Williams, "Life Cycle Water Use of Low-Carbon Transport Fuels," *Energy Policy*, vol. 38 (2010), pp. 4933-4944).

[54] For more information, see CRS Report R41460, *Cellulosic Ethanol: Feedstocks, Conversion Technologies, Economics, and Policy Options*, by Randy Schnepf.

[55] For example, dry cooling is not equally effective for power plant cooling in locations with temperatures above 100°F, or substituting municipal wastewater may not be an option for remote facilities.

[56] J. Leshy, "Notes on a Progressive National Water Policy," *Harvard Law & Policy Review*, vol. 3, no. 1 (Winter 2009), pp. 133-159.

[57] D. H. Getches, "The Metamorphosis of Western Water Policy: Have Federal Laws and Local Decisions Eclipsed the States' Role?," *Stanford Environmental Law Journal*, vol. 20 (2001), pp. 3-72.

[58] For more on rare minerals in the energy sector, CRS Report R41347, *Rare Earth Elements: The Global Supply Chain*, by Marc Humphries.

[59] Elcock 2010.

[60] Elcock 2010 considered that only energy crops would generate an increase in water consumption. The study's projections for water consumption were based on projections for different biofuels, including corn-based ethanol, cellulosic ethanol, and biodiesel.

[61] King 2010, p. 1162.

[62] NETL, *Estimating Freshwater Needs to Meet Future Thermoelectric Generation Requirements*, 2007.

[63] Elcock 2010, p. 451. For more, see CRS Report RL34738, *Cellulosic Biofuels: Analysis of Policy Issues for Congress*, by Kelsi Bracmort et al.

[64] Argonne National Laboratory released for comment the following report: ANL, *DRAFT Water Use in the Development and Operation of Geothermal Power Plants*, September 2010, http://www1.eere.energy.gov/geothermal/ pdfs/geothermal_water_use_draft.pdf, hereafter referred to as Argonne 2010. The draft links its geothermal water use findings with water intensities of other electricity generation. The report cited references from the early 1990s for fuel processing, and those references cited materials from the late 1970s to the early 1990s. Other articles also use data from these sources (e.g., V. Fthenakis and H.C. Kim, "Life-cycle Uses of Water in U.S. Electricity Generation," *Renewable and Sustainable Energy Reviews*, vol. 14 (2010), pp. 2039 2048). Continued reliance on older sources demonstrates a knowledge gap on current water use in fuel acquisition and preparation. Changes in the U.S. fossil fuel practices (e.g., natural gas extraction through fracturing) in recent decades may have significantly altered water use for fuel.

[65] Chemicals added to the water at a thermoelectric facility to extend equipment life and improve operational efficiency (e.g., demineralize regenerants and rinses to prevent biological growth) can degrade water quality when discharged into the environment. To comply with water quality regulations promulgated under the Clean Water Act, thermoelectric facilities constructed after January 17, 2002, generally use evaporative cooling. EPA has begun revising these rules, which were issued in 1982. Separately under the Clean Water Act, EPA regulates cooling water intake structures, which protect fish from entrainment at the intake

point. EPA is developing a rule to address water intake structures for some existing thermoelectric facilities, which is expected to be proposed in 2011; the rule may affect decisions about power plan retirements and cooling technology.

66 DOE, *Concentrating Solar Power Commercial Application Study: Reducing Water Consumption of Concentrating Solar Power Electricity Generation*, no date,: http://www1.eere.energy.gov/solar/pdfs/csp_water_study.pdf.

67 The Hewlett Foundation and The Energy Foundation, *The Last Straw: Water Use by Power Plants in the Arid West*, April 2003, pg. 12, http://www.catf.us/publications /reports/The_Last_Straw.pdf. Another source estimates that dry cooling and hybrid wet-dry cooling have capital costs 1.1 to 5 times higher than wet cooling (California Energy Commission, *Comparison of Alternate Cooling Technologies for California Power Plants: Economic, Environmental and Other Tradeoffs*, Consultant Report 500-02-079F, Sacramento, CA, Feb. 2002, p. 1-9, http://www.energy.ca.gov/ reports/2002-07-09_500-02-079F.PDF). This same source provides an example of the cooling system capital costs for a 500 MW combined-cycle power plant — $3.6 million for a wet cooling tower versus $25.5 million for a dry cooling system (ibid., p. 5-16 and p. 5-39).

68 Letter from Ric O'Connell, Renewable Energy Consultant, Black & Veatch, to Environmental Working Group, Renewable Energy Transmission Initiative, June 25, 2008.

69 The loss in generation efficiency is from insufficient cooling of the turbine exhaust steam, increasing steam turbine back pressure. While efficiency losses also are typical of wet-cooled systems when inlet water temperatures exceed design temperatures, the lower cooling capacity of air versus water makes dry cooling more sensitive to temperature changes and efficiency losses than wet cooling. A power plant with dry cooling can experience a 1% loss in efficiency for each 1°F increase of the condenser, which is limited by ambient temperatures (C. Kutscher, A. Buys, and C. Gladden, *Hybrid Wet/Dry Cooling for Power Plants*, NREL, presented at Parabolic Trough Technology Workshop, Incline Village, NV, Feb. 14, 2006, http://www.nrel.gov/docs/fy06osti/40026.pdf.). Electricity output is also decreased by the additional electricity requirement of running the fans and pumps of the air cooling system.

70 While there is interest in using alternative water sources such as coastal waters for cooling, there is concern about harm to marine ecosystems.

71 For example, the Palo Verde Nuclear Generating Station near Phoenix, AZ, uses wastewater effluent for cooling.

72 D. Lawrence et al., *Power Plant Engineering – Design and Construction*, (Black & Veatch, Chapman & Hall: New York, 1996).

73 J. A. Veil., J. M. Kupar, and M. G. Puder, Argonne National Laboratory, *Use of Mine Pool Water for Power Plant Cooling*, September 2003, http://www.ipd.anl.gov /anlpubs /2006/11/57830.pdf.

74 Whether hydroelectric power generation is a renewable is a subject of debate largely because of the environmental impacts of dams and their reservoirs. For more information on small-scale and low-head hydropower, see CRS Report R41089, *Small Hydro and Low-Head Hydro Power Technologies and Prospects*, by Richard J. Campbell.

75 EPRI, Assessment of Waterpower Potential and Development Needs (Palo Alto, CA: 2007), http://mydocs.epri.com/ docs/public/000000000001014762.pdf.

76 Idaho National Laboratory, *Feasibility Assessment of the Water Energy Resources of the United States for New Low Power and Small Hydro Classes of Hydroelectric Plants*, (DOE, Washington, DC: 2006), http://hydropower.inel.gov/ resourceassessment/pdfs /main_

report_appendix_a_final.pdf. This study estimated preliminarily feasible opportunities for increasing U.S. hydroelectric generation with a total potential of 30 MW.

[77] DOE, *National Electric Transmission Congestion Study*, December 2009, http://congestion 09.anl.gov/documents/ docs/Congestion_Study_2009.pdf, presents areas where renewable energy development using existing technologies is constrained by transmission, and other areas where development would be constrained if technologies mature (e.g., offshore wind). The areas renewable energy transmission constrained areas are shown in the report on page 23.

[78] Intermittent electricity sources can be used to meet electricity demand for activities that can be ramped up or down (e.g., hydrogen generation via electrolysis, though this technology is still developing commercially) and to reduce generation from other sources when they are available.

[79] DOE, How an Enhanced Geothermal System Works, Washington, DC, March 31, 2006, http://www1.eere.energy.gov/geothermal/printable_versions/egs_animation.html

[80] USGS, *Substantial Power Generation from Domestic Geothermal Resources*, Reston, VA, Sept. 29, 2008, http://www.usgs.gov/newsroom/article.asp?ID=2027&from=rss_home, pg. 5.

[81] For examples of a geothermal potential map, see the following websites: http://www.nrel.gov /gis/images/geothermal_resource2009-final.jpg, and http://www.azgs.state /geothermal_ 6b.jpg.

[82] B.D. Green and G. Nix, *Geothermal - Energy Under Our Feet*, NREL, NREL/TP-840-40665, Nov. 2008.

In: The Energy-Water Nexus: Key Issues ISBN: 978-1-62417-753-8
Editors: Tucker Webb and Ellis Porter © 2013 Nova Science Publishers, Inc.

Chapter 3

TESTIMONY OF DR. MICHAEL WEBBER, ASSOCIATE DIRECTOR, CENTER FOR INTERNATIONAL ENERGY AND ENVIRONMENTAL POLICY, THE UNIVERSITY OF TEXAS AT AUSTIN. HEARING ON THE "ENERGY-WATER NEXUS"*

Mr. Chairman and Members of the Committee, thank you so much for the invitation to speak before your committee on the nexus of energy and water. My name is Michael Webber, and I am the Associate Director of the Center for International Energy and Environmental Policy and Assistant Professor of Mechanical Engineering at the University of Texas at Austin. I appear here today to share with you my perspective on important trends and policy issues related to this nexus.

My testimony today will make four main points:

1) Energy and water are interrelated,
2) The energy-water relationship is already under strain,
3) Trends imply these strains will be exacerbated, and
4) There are different policy actions that can help.

I will briefly elaborate on each of these points during this testimony.

---

* This is an edited, reformatted and augmented version of testimony given on March 10, 2009 before the Senate Committee on Energy and Natural Resources.

ENERGY AND WATER ARE INTERRELATED

Energy and water are interrelated: we use energy for water, and we use water for energy.

For example, we use energy to heat, treat and move water. Water heating alone is responsible for 9% of residential electricity consumption in the U.S. And, nationwide, water and wastewater treatment and distribution combined require about 3% of the nation's electricity. However, regionally, that number can be much higher. In California, where water is moved hundreds of miles across two mountain ranges, water is responsible for approximately 15% of the state's total electricity consumption. Similarly large investments of energy for water occurs wherever water is scarce and energy is available.

In addition to using energy for water, we also use water for energy. We use water directly through hydroelectric power generation at major dams, indirectly as a coolant for thermoelectric power plants, and as a critical input for the production of biofuels. The thermoelectric power sector—comprised of power plants that use heat to generate power, including those that operate on nuclear, coal, natural gas or biomass fuels—is the single largest user of water in the United States. Cooling of power plants is responsible for the withdrawal of nearly 200 billion gallons of water per day. This use accounts for 49% of all water withdrawals in the nation when including saline withdrawals, and 39% of all freshwater withdrawals, which is about the same as for agriculture. On average, anywhere between 1 to 40 gallons of water is needed for cooling for every kilowatt-hour of electricity that is generated. However, while power plants withdraw vast amounts of water, very little of that water is actually consumed; most of the water is returned to the source though at a different temperature and with a different quality. Thus, while power plants are major users of water, they are not major consumers of water, which is in contrast with the agriculture sector, which consumes all the water it withdraws.

THE ENERGY-WATER RELATIONSHIP IS ALREADY UNDER STRAIN

Unfortunately, the energy-water relationship introduces vulnerabilities whereby constraints of one resource introduces constraints in the other. For example, during the heat wave in France in 2003 that was responsible for approximately 10,000 deaths, nuclear power plants in France had to reduce their power output because of the high inlet temperatures of the cooling water. Environmental regulations in France (and the United States) limit the rejection temperature of power plant cooling water to avoid ecosystem damage from thermal pollution (e.g. to avoid cooking the plants and animals in the waterway). When the heat wave raised river temperatures, the nuclear power plants could not achieve sufficient cooling within the environmental limits, and so they reduced their power output at a time when electricity demand was spiking by residents turning on their air conditioners. In this case, a water resource constraint became an energy constraints.

In addition to heat waves, droughts can also strain the energy-water relationship. During the drought in the southeastern United States in early 2008, nuclear power plants were within weeks of shutting down because of limited water supplies. Today in the west, a severe multi-year drought has lowered water levels behind Hoover Dam, introducing the risk that Las Vegas will lose a substantial portion of its drinking water at the same time the dam's hydroelectric turbines quit spinning, which would cut off a significant source of power for Los Angeles. In addition, power outages hamper the ability for the water/wastewater sector to treat and distribute water. Thus, strain in the energy-water nexus is very real in the United States and is here today.

It is important to note that while constraints in one resource introduce constraints on the other, the corollary of that relationship is also true. That is, both resources can be enabling for the other: with unlimited energy, we could have unlimited freshwater; with unlimited water, we could have unlimited energy.

Trends Imply These Strains Will Be Exacerbated

While the energy-water relationship is already under strain today, trends imply that the strain will be exacerbated unless we take appropriate action. There are four key pieces to this overall trend:

1) Population growth, which drives up total demand for energy and water,
2) Economic growth, which can drive up per capita demand for both energy and water,
3) Climate change, which intensifies the hydrological cycle, and
4) Policy choices, whereby we are choosing to move towards more energy-intensive water and more water-intensive energy.

Population Growth Will Put Upward Pressure on Demand for Energy and Water

Population growth over the next few decades might yield another 100 million people in the United States over the next four decades, each of whom will need energy and water to survive and prosper. This fundamental demographic trend puts upward pressure on demand for both resources, thereby potentially straining the energy-water relationship further.

Economic Growth Will Put Upward Pressure on Per Capita Demand for Energy and Water

On top of underlying trends for population growth is an expectation for economic growth. Because personal energy and water consumption tend to increase with affluence, there is the risk that the per capita demand for energy and water will increase due to economic growth. For example, as people become wealthier they tend to eat more meat (which is very water intensive), and use more energy and water to air condition large homes or irrigate their lawns. Also, as societies become richer, they often demand better environmental conditions, which implies they will spend more energy on wastewater treatment. However, it's important to note that the use of efficiency and conservation measures can occur alongside economic growth, thereby counteracting the nominal trend for increased per capita consumption of energy and water. At this point, looking forward, it is not clear whether technology, efficiency and conservation will continue to mitigate the upward pressure on per capita consumption that are a consequence of economic growth. Thus, it's possible that the United States will have a compounding effect of increased consumption per person on top of a growing number of people.

Climate Change Is Likely to Intensify Hydrological Cycles

One of the important ways climate change will manifest itself it through an intensification of the global hydrological cycle. This intensification is likely to mean more frequent and severe droughts and floods along with distorted snowmelt patterns. Because of these changes to the natural water system, it is likely we will need to spend more energy storing, moving, treating and producing water. For example, as droughts strain existing water supplies, cities might consider production from deeper aquifers, poorer-quality sources that require desalination, or long-haul pipelines to get the water to its final destination. Las Vegas, San Diego and Dallas are already considering some version of these options, all of which are extremely energy-intensive. Desalination in particular is alarming because it is approximately ten times more energy-intensive than production from surface freshwater sources such as rivers and lakes. Some areas are considering a combination of desalination plus long-haul pipelines, which has a compounding effect for energy use.

Policy Choices Exacerbate Strain in the Energy-Water Nexus

On top of the prior three trends is a policy-driven movement towards more energy-intensive water and water-intensive energy.

We are moving towards more energy-intensive water because of increasingly strict treatment standards for water and wastewater, which requires more energy than traditional approaches that met prior standards. In addition, instead of a push for water efficiency and conservation, many municipalities are pushing for new supplies of water starting with sources that are farther away and lower quality, and thereby require more energy to get them to the right quality and location.

For a variety of reasons, including the desire to produce a higher proportion of our energy from domestic sources and to decarbonize our energy system, many of our preferred energy choices are more water-intensive. For example, nuclear energy is produced domestically, but is also more water-intensive than other forms of power generation. The move towards more water-intensive energy is especially relevant for transportation fuels such as unconventional fossil fuels (oil shale, coal-to-liquids, gas-to-liquids, tar sands), electricity, hydrogen, and biofuels, all of which can require significantly more water to produce than gasoline (depending on how you produce them). It is important to note that the push for renewable electricity also includes solar photovoltaics and wind power, which require very little water, and so not all future energy choices are worse from a water-perspective.

Almost all unconventional fossil fuels are more water-intensive than domestic, conventional gasoline production. While gasoline might require a few gallons of water for every gallon of fuel that is produced, the unconventional fossil sources are typically a few times more water-intensive. Electricity for plug-in hybrid electric vehicles (PHEVs) or electric vehicles

(EVs) are appealing because they are clean at the vehicle's end-use and it's easier to scrub emissions at hundreds of smokestacks millions of tailpipes. However, powerplants use a lot of cooling water, and consequently electricity can also be about twice as water-intensive than gasoline per mile traveled if the electricity is generated from the standard U.S. grid. If that electricity is generated from wind or other water-free sources, then it will be less water-consumptive than gasoline. Hydrogen can also be more water-intensive than gasoline, depending on how it is produced. If made from steam methane reforming or electrolysis from water-free electrical sources such as wind, then hydrogen is no worse than gasoline (and potentially much better). However, if hydrogen is made from electrolysis using electricity from the standard U.S. grid, then producing hydrogen might consume more than 25 gallons of water and withdraw more than 1000 gallons for every gallon of gasoline equivalent energy that is produced. Though unconventional fossil fuels, electricity and hydrogen are all potentially more water-intensive than conventional gasoline by up to a factor of 10 or so, biofuels are particularly water-intensive. Growing biofuels consumes more than 1000 gallons of water for every gallon of fuel that is produced. Sometimes this water is provided naturally from rainfall, however for a non-trivial proportion of our biofuels production, irrigation is used. Irrigated biofuels from corn or soy can consume twenty or more gallons of water for every mile traveled.

Note that for the sake of analysis and regulation, it is convenient to consider the water requirements per mile traveled. Doing so incorporates the energy density of the final fuels plus the efficiency of the engines, motors or fuel cells with which they are compatible.

If we compare the water requirements per mile traveled with projections for future transportation miles and combined those figures with mandates for the use of new fuels, such as biofuels, the water impacts are startling. Water consumption might go up from approximately one trillion gallons of water per year to make gasoline (with ethanol as an oxygenate), to a few trillion gallons of water per year. To put this water consumption into context, each year the United States consumes about 36 trillion gallons of water. Consequently, it is possible that water consumption for transportation will more than double from less than 3% of national use to more than 7% of national use. In a time when we are already facing water constraints, it is not clear we have the water to pursue this path. Essentially we are deciding to switch from foreign oil to domestic water for our transportation fuels, and while that might be a good decision for strategic purposes, I advise that we first make sure we have the water.

THERE ARE DIFFERENT POLICY ACTIONS THAT CAN HELP

Because there are many rivers, watersheds, basins and aquifers that span several states and/or countries, there is a need for federal engagement on energy-water issues.

Unfortunately, there are some policy pitfalls at the energy-water nexus. For example, energy and water policymaking are disaggregated. The funding and oversight mechanisms are separate, and there are a multitude of agencies, committees, and so forth, none of which have clear authority. It is not unusual for water planners to assume they have all the energy they need and for energy planners to assume they have the water they need. If their assumptions break down, it could cause significant problems. In addition, the hierarchy of policymaking is dissimilar. Energy policy is formulated in a top-down approach, with powerful federal energy agencies, while water policy is formulated in a bottom-up approach, with powerful local and state water agencies. Furthermore, the data on water quantity are sparse, error-prone, and inconsistent. The United States Geological Survey (USGS) conducted its last survey on water consumption in 1995 and its last published data on water withdrawals are from 2000. National databases of water use for power plants contain errors, possibly due to differences in the units, format and definitions between state and federal reporting requirements. For example, the definitions for water use, withdrawal and consumption are not always clear. And, water planners in the east use "gallons" and water planners in the west use "acre-feet," introducing additional risk for confusion or mistakes.

Despite the potential pitfalls, there are policy opportunities at the energy water nexus. For example, water conservation and energy conservation are synonymous. Policies that promote water conservation also achieve energy conservation. Policies that promote energy conservation also achieve water conservation. It is my opinion that robust energy and water policies should begin with conservation because of the cascading cross-over benefits they offer.

Thankfully, the federal government has some effective policy levers at its disposal. I recommend the following policy actions for the energy-water nexus:

1) *Collect, maintain and make available accurate, updated and comprehensive water data, possibly through the USGS.* The Department of Energy's Energy Information Administration maintains an extensive database of accurate, up-to-date and comprehensive information on energy production, consumption, trade, and price available with temporal and geographic resolution and standardized units. Unfortunately, there is no equivalent set of data for water. Consequently, analysts, policymakers and planners lack suitable data to make informed decisions.
2) *Establish federal oversight for water quantity.* The Environmental Protection Agency has oversight of water quality, but it's not clear if any agency has oversight of water quantity.
3) *Establish strict standards in building codes for water efficiency.* Building codes should include revised standards for low-flow appliances, water-heating efficiency, purple-piping for reclaimed water, rain barrels and so forth in order to reduce both water and energy consumption.
4) *Invest heavily in water-related R and D to match recent increases in energy-related R and D.* R and D investments are an excellent policy option for the federal government because state/local governments and industry usually are not in a position to adequately invest in research. Consequently, the amount of R and D in the water sector is much lower than for other sectors such as pharmaceuticals, technology, or energy. Furthermore, since energy-related R and D is expected to go through a surge in funding, it would be appropriate from the perspective of the energy-water nexus to raise water-related R and D in a commensurate way. Topics for R and D include low-energy water treatment, novel approaches to desalination, remote leak detectors for water infrastructure, and air-cooling systems for power plants. In addition, DoE's R and D program for biofuels should emphasize feedstocks such as cellulosic sources or algae that do note require freshwater irrigation.
5) *Support the use of reclaimed water at powerplants, industry and agriculture.* Using reclaimed water for powerplants, industry and agriculture spares a significant amount of energy. However there are financing, regulatory and permitting hurdles in place that restrict this option.
6) *Rethink water markets.* Water is widely expected to be free and unlimited, even though water is a limited resource that we should

value highly. Consequently, it is worthwhile to consider implementing water markets that balance our competing needs to meet our social justice and human rights goals (that is, everyone needs water to survive, whether rich or poor), while also meeting our need to discourage water waste through high prices. Block pricing, whereby the first amount of water usage is cheap or free in order to meet our survival needs, after which the price escalates significantly in order to curtail water use for non-critical purposes, might be a fruitful approach.

The energy-water nexus is a complicated, important issue, and so I am very pleased to know that you are being attentive to the matter.

Mr. Chairman, that concludes my testimony. I'll be pleased to answer questions at the appropriate time.

In: The Energy-Water Nexus: Key Issues ISBN: 978-1-62417-753-8
Editors: Tucker Webb and Ellis Porter © 2013 Nova Science Publishers, Inc.

Chapter 4

ENERGY AND WATER: CONNECTION AND CONFLICT*

House Committee on Natural Resources, Democratic Staff

Note: This report has not been officially adopted by the Committee on Natural Resources and may not necessarily reflect the views of its Members.

* This is an edited, reformatted and augmented version of House Committee on Natural Resources, Democratic Staff Publication, dated August 17, 2012.

ENERGY AND WATER: CONNECTION AND CONFLICT

Energy and water are two fundamental building blocks of the U.S. economy. Power plants need water to operate. Electric water pumps move water around the country, especially in the West. Because energy and water are inextricably linked, they are in a constant interplay of supply and demand where a change in the availability one resource can impact the usage of the other and vice versa.

Especially during droughts, water scarcity can cause conflict as drinking and agriculture needs increasingly compete against energy needs.

ENERGY'S THIRST FOR WATER

Energy's thirst for water manifests itself in several ways. To produce electricity in nuclear and fossil-fueled power plants requires substantial volumes of water. Extracting and processing raw materials such as coal and natural gas requires water, but doing so also risks polluting and contaminating surface and groundwater sources. This not only could jeopardize drinking water supplies, but could also undermine future energy production if current water sources become scarce or rendered unusable. The nation's energy system is one of the largest consumers of water, a relationship that cannot be sustained into the future as the U.S. population grows and the climate changes.

ELECTRICITY PRODUCTION

- In the United States, nuclear and fossil-fueled power plants *withdraw* 143 billion gallons of freshwater[1] every day from rivers, lakes, stream, and aquifers. This is roughly equivalent to the daily domestic water usage of 140 New York Cities[2] or the daily water needed to irrigate the fields of nearly 17 Nebraskas.
- These same plants *consume* 2.2 to 5.9 billion gallons of water every day[3], primarily through evaporation of steam generated to turn their turbines. While substantially less than the amount withdrawn, consumed water is a problem in areas where water supplies are already stretched since this water is lost to other users in the region such as farmers.

- U.S. coal plants alone were responsible for 67% of the water withdrawals and 65% of the consumption in 2008[4]. However, nuclear power plants require the most water on a per plant basis.
- Since 2004, water supply problems have led at least a dozen power plants to temporarily reduce their power output or shut down entirely and prompted at least eight states to deny new plant proposals.[5]

EXTRACTION AND REFINING

- Mountaintop removal mining has buried almost 2,000 miles of Appalachian streams, contaminating drinking water and destroying aquatic ecosystems[6].
- Producing uranium for nuclear power plants has contaminated U.S. surface and groundwater supplies in 14 states[7].
- The overall water needs for hydraulic fracturing and shale mining are still unknown. Their needs may challenge supplies and infrastructure, especially in parts of the West where water supplies are already limited. Nonetheless, companies have acquired rights to billions of gallons of water to remove oil from Western Colorado's tar shale, setting up future conflicts with other important water users such as farmers and municipalities[8].

FUTURE TRENDS

- By 2035, U.S. primary energy consumption is projected to increase an additional 9% over 2010 levels[9], which will also require additional water use.
- Over 85% of our electric power is still projected to come from nuclear and fossil-fueled power plants in 2035.[10]
- In addition, water-intensive oil and natural gas production from unconventional sources in the United States is increasing rapidly due to technological advances in the industry.
- The energy sector is the fastest-growing water consumer in the United States. Studies predict that the energy sector will be responsible for 85% of the growth in water consumption between 2005 and 2030.[11]

- To meet future energy demand, the United States needs to ensure that water resources are easily accessible and reliable.

Global Warming Threatens U.S. Energy Supply

The energy-water connection directly intersects with global warming in two specific ways. First, the extraction, treatment, and distribution of water requires energy which in turn produces global warming pollution. Second, global warming is already impacting sea level, snowpack, rainfall patterns, and drought. These increasingly challenging conditions will constrain water resources, putting future energy supply at risk.

Energy's Impact on Climate

- The combustion of fossil fuels to generate electricity is the largest single source of CO_2 emissions in the nation, accounting for nearly 40% of total 2010 CO_2 emissions, which is equivalent to about 2,258 million metric tons[12] or approximately the CO_2 emitted by 513 million cars in a year[13].
- Coal-fired power plants are the largest source of global warming pollution in the U.S. electric power sector. Even newer coal plants produce more than double the amount of carbon emissions of a new natural gas combined cycle plant (2,464 compared to 1,076 lbCO_2e/MWh).[14] U.S. coal accounts for approximately 70% of CO_2 emissions from U.S. power plants and 34% of CO_2 emissions from total U.S. fossil fuel energy use.
- Methane gas that escapes from natural gas fields, also referred to as fugitive gas emissions, are also a concern given that methane has over 23 times the global warming impact of CO_2.

  Studies by scientists at the National Oceanic and Atmospheric Administration (NOAA) indicate that natural gas drilling might generate twice as much global warming pollution as previously estimated.[15]

CLIMATE'S IMPACT ON ENERGY

- Based on observational records, temperatures have warmed roughly 1.33°F (0.74°C) during the last century averaged over all land and ocean surfaces, according to the Intergovernmental Panel on Climate Change. More than half of this warming—about 0.72°F (0.4°C)—has occurred since 1979. Because oceans tend to warm and cool more slowly than land areas, continents have warmed the most (about 1.26° F or 0.7° C since 1979), especially over the Northern Hemisphere[16].
- According to NASA records, the first 11 years of the 21st century experienced notably higher temperatures compared to the middle and late 20th century. Nine of the top 10 warmest years globally have occurred since 2000.
- Based on observational data, the five fastest warming states — Rhode Island, Massachusetts, New Jersey, Arizona and Maine — warmed at rates twice the continental United States average of 0.127°F per decade over the last 100-year period.[17]
- The IPCC (International Panel for Climate Change) predict a global rise of between 1.1°C and 6.4°C (1.98°F – 11.52°F)[18] by 2100. The best estimate of global temperature increase is between 1.8°C and 4°C (3.24°F – 7.2°F).[19]
- Due to temperature increases and changes in precipitation, more frequent droughts in most of the continental United States are expected by 2050.[20]
- Water scarcity, lower summer river flows, and higher river water temperatures due to global warming could lead to a 4.4 to16% decrease in power plant capacity by 2060[21].
- This year's extreme weather may be a preview into global warming. NOAA reported that for the continental United States, the August 2011-July 2012 period was the warmest 12-month period since record keeping began in 1895[22].

  This broke the record set in June 2012 which in turn broke the record in May 2012. The past fourteen months have featured the United State's 2nd warmest summer, 4th warmest winter, and warmest spring on record[23]. Every state in the lower 48 states had warmer than average temperatures for the period, except Washington, which was near normal.[24]
- In July 2012, almost 65% of the continental United States was experiencing drought[25], and the country has also been plagued by

strong storms and devastating wildfires, which are consistent with scientific predictions of the impacts of global warming[26].

- According to an August 2012 study published in the journal *Nature Climate Change,* observations now confirm what climate models have previously predicted: Drought will become more severe in the United States in the coming decades, and climate change will amplify drought frequency and intensity[27].

REGIONAL IMPACTS

Taking into account projected increases in population, power generation, and water demand, as well as climate change forecasts of temperature and available precipitation, 14 states face an extreme or high risk to water sustainability, with limitations on use expected as demand exceeds supply by 2050.[28]

The Southwest, Great Plains, and Southeast will be particularly challenged.

THE WEST/SOUTHWEST[29]

- Electricity generating capacity throughout the West and Northwestern Canada is expected to grow by 22% between 2010 and 2035.[30]
- Large hydroelectric facilities including the Hoover and Glen Canyon Dams depend on reservoirs that have recently had record-low water levels due to extensive droughts,[31,32] leading to substantially reduced electricity output.
- Rapidly growing cities have been forced during extended drought and record-breaking heat to revert to unsustainable groundwater pumping in order to meet municipal water needs.[33]

  Water supply conflicts involving these cities are considered highly likely by 2025[34], particularly since many share the same water resources from the Colorado Basin.

THE GREAT PLAINS[35]

- Electricity generating capacity in Texas is expected to grow by 19% between 2010 and 2035 and by 7% in Nebraska, Kansas, Oklahoma, and parts of Missouri, Arkansas, Louisiana, and New Mexico.[36]
- The agriculture sector, which is the energy sector's largest competitor for water resources, is an important part of the Great Plains states' economies. States will have to make tough decisions regarding water prioritization.
- Water demand for hydraulic fracturing in the Eagle Ford formation will increase 10-fold by 2020, and double again by 2030[37]. A single company that currently has 10 wells in the area is planning to drill another 200 in 2012, increasing the water demand the equivalent of the water usage for about 50,000 people for one year[38]. Since agriculture currently uses approximately 60% of water, the additional demand for water for hydraulic fracturing may exacerbate tensions between farmers and the energy sector.
- Given the 2012 drought, water is starting to become an issue for the oil companies in Harper County Oklahoma with restrictions on water withdrawals from some streams starting in early July. Because there is not much groundwater available companies must seek privately owned supplies from farmers and adjacent municipalities. The price of water has gone up to as much as 25 cents a barrel (1 barrel is 42 gallons)[39].

THE SOUTHEAST[40]

- Electricity generating capacity throughout much of the Southeast is expected to grow by 12% between 2010 and 2035.[41]
- Coal and nuclear power plants that rely on once-through cooling are responsible for 66% of freshwater withdrawals in the Southeast.[42] Plants have to cut output or sometimes shutdown completely if water levels become too low for withdrawal, the water is too warm to cool the plant, or the discharged water is too warm to return to its source.
- During prolonged heat in the summer of 2010, Browns Ferry nuclear plant in Alabama had to cut power by 50% for five consecutive weeks.[43]

- During the 2007-08 drought, South Carolina sued North Carolina over a plan by certain North Carolina cities to withdraw water from the Catawba River.[44]

SOLUTIONS

Although the interconnection between energy and water creates challenges for the United States, there are near-term solutions and technologies we can utilize now that will enhance the efficient use of our water and energy resource. These include:

- *Increasing development of water-free energy technologies such as renewables:* Wind and photovoltaic solar electricity generation require practically zero water. Solar, wind and other forms of renewable energy offer an alternative to fossil fuel and nuclear power plants that require large amounts of water both to generate electricity and to extract uranium, coal and natural gas.
- *Increasing efficiency of energy and water systems, stretch local water supplies:* Fixing leaks in public drinking water systems prevents water loss and saves energy required to move it through the system. We can also maximize our existing resources by stretching local water supplies through water recycling.
- *Exploring alternative water sources for electricity generation:* Recycled municipal wastewater is a reliable water source that is available in relative abundance across the United States. Other alternative water sources include waste water from oil and gas wells, mine pool water, and industrial process water.
- *Innovatively financing the solutions:* Extend the federal tax credits for new water-free renewable energy projects. A federal renewable electricity standard (RES) will also help renewables flourish.
- *Encouraging investment in innovation:* Technology can help save water and energy in a myriad of ways, from the appliances we use, to the homes and offices we live and work in, to the cars and trucks we drive. As we invest in upgrading our aging energy and water infrastructure, we must incentivize public and private investment in smart sensors, new optimization software, highly efficient pumps, and other innovative technology. Efficiency will stretch our existing

energy and water resources at a fraction of the cost of expensive new power plants, pipes, pumping stations, and water treatment plants.

End Notes

[1] USGS (2009). *Estimated Use of Water in the United States in 2005*. United States Geological Survey (USGS Circular 1344).

[2] UCS (2010). *The Energy-Water Collision: 10 Things You Should Know*. Union of Concerned Scientists.

[3] UCS (2011a). *Freshwater Use by U.S. Power Plants: Electricity's Thirst for a Precious Resource*. Union of Concerned Scientists.

[4] *Ibid.*

[5] B.K. Sovacool (2009). Running on empty: The electricity-water nexus and the US electric utility sector. *Energy Law Journal, 30*(1), 11-51.

[6] EPA (2010). EPA issues comprehensive guidance to protect Appalachian communities from harmful environmental impacts of mountain top mining. Press release, April 1.

[7] World Information Service on Energy (2012). *Decommissioning Projects – USA*. Retrieved from http://www.wise-uranium.org/udusa.html

[8] S. Lipsher (2008). *Colorado's oil shale draws Shell's interest*. Retrieved from http://www.chron.com/business/energy/article/Colorado-s-oilshale-draws-Shell-s-interest-1785278.php

[9] EIA (2012c). *Annual Energy Outlook 2012, Table A2: Energy consumption by sector and source*. Energy Information Administration.

[10] EIA (2012d). *Annual Energy Outlook 2012, Table A8: Electricity supply, disposition, prices, and emissions*. Energy Information Administration.

[11] N. Carter (2011). *Energy's Water Demand: Trends, Vulnerabilities, and Management*. Congressional Research Service (CRS Publication No. R-41507).

[12] EPA (2012). *Inventory of U.S. Greenhouse Gas Emissions and Sinks: 1990 – 2010*. Environmental Protection Agency.

[13] A.H. Rosenfeld & S. Kumar (2001). *Tables to Convert Energy or CO2 to Familiar Equivalents*. Retrieved from http://www.energy.ca.gov/commissioners /rosenfeld_docs /Equivalence-Matrix_2001-05.pdf

[14] T. Skone (2012). *Life Cycle Water Withdrawal and Consumption in the Thermoelectric Power Industry*. DOE National Energy Technology Laboratory. Presented to the U.S. House of Representatives, May 30, 2012.

[15] E. Shogren (2012). *Fracking's Methane Trail: A Detective Story*. National Public Radio. Retrieved from http:*//www.npr.org/2012/05/17/151545578/frackings-methane-trail-a-detective-story*

[16] https://www2.ucar.edu/faq/very-frequently-asked

[17] C. Tebaldi et al. (2012). *The Heat Is On: U.S. Temperature Trends*. Climate Central. Retrieved from http://www.climatecentral.org/wgts/heatis-on/HeatIsOnReport.pdf

[18] Meehl *et al.*, Chap. 10: Global Climate Projections, Sec. 10.ES: Mean Temperature, in IPCC AR4 WG1 2007.

[19] Meehl *et al.*, Chap. 10: Global Climate Projections, Sec. 10.ES: Mean Temperature, in IPCC AR4 WG1 2007.

[20] K. Strzepek et al. (2010). Characterizing changes in drought risk for the United States from climate change. *Environmental Research Letters, 5*(4), 1-9.

[21] M. van Vliet et al. (2012). Vulnerability of US and European electricity supply to climate change. *Nature Climate Change,* Vol. advance online publication (03 June 2012), doi:10.1038/nclimate1546

[22] http://www.ncdc.noaa.gov/sotc/national/2012/7

[23] http://www.ncdc.noaa.gov/sotc/national/2012/7

[24] http://www.ncdc.noaa.gov/sotc/national/2012/7

[25] http://droughtmonitor.unl.edu/

[26] http://epw.senate.gov/public/index.cfm?FuseAction=Hearings.Testimony&Hearing_ID =c0293 eca-802a-23ad-4706-02abdbf7f7c3&Witness_ID=cba9c00e-6d03-4da7-a87e-7080bfdbb3bd

[27] http://www.eenews.net/climatewire/2012/08/15/9,

[28] S. Roy et al. (2010). *Evaluating Sustainability of Projected Water Demands under Future Climate Change Scenarios.* Tetra Tech, Inc. Retrieved from http://rd.tetratech.com /climatechange/projects/doc/Tetra_Tech_Climate_Report_2010_lowres.pdf

[29] Includes Arizona, Colorado, Nevada, New Mexico, and Utah

[30] EIA (2012e). *Annual Energy Outlook 2012, Table 57: Electricity Generating Capacity by Electricity Market Module Region and Source*. Energy Information Administration.

[31] U.S. Bureau of Reclamation. *Lake Mead at Hoover Dam, elevation (feet).* Retrieved from http://www.usbr.gov/lc/region/g4000/hourly/meadelv.html

[32] U.S. Bureau of Reclamation. *Upper Colorado Region water operations: Current Status: Lake Powell.* Retrieved from http://www.usbr.gov/uc/water/crsp/cs/gcd.html

[33] T. Hutchins-Cabibi et al. (2006). *Water in the urban Southwest: An updated analysis of water use in Albuquerque, Las Vegas, and Tucson*. Western Resource Advocates.

[34] UCS (2011b). *The Energy-Water Collision: Power and Water at Risk.* Union of Concerned Scientists.

[35] Includes Arkansas, Kansas, Nebraska, Oklahoma, and Texas

[36] EIA (2012e).

[37] http://www.bloomberg.com/news/2011-06-13/worst-drought-in-more-than-a-century-threa tens-texas-oil-natural-gas-boom.html

[38] http://www.bloomberg.com/news/2011-06-13/worst-drought-in-more-than-a-century-threatens -texas-oil-natural-gas-boom.html

[39] http://www.eenews.net/public/energywire/2012/07/30/1; Dai (2012) Increasing drought under global warming in observations and models, Nature Climate Change, doi: 10.1038/nclimate1633

[40] Includes Alabama, Florida, Georgia, Kentucky, Mississippi, North Carolina, South Carolina, Tennessee, and Virginia

[41] EIA (2012e).

[42] UCS (2011b).

[43] NRC (2010). *Power Reactor Status Report for July 27, 2010*. Nuclear Regulatory Commission. Retrieved from http://www.nrc.gov/readingrm/doc-collections/event-status/reactor-status /2010/20100727ps.html

[44] USA Today (2008). *Drought spreading in Southeast*. Retrieved from http://www.usatoday.com /weather/drought/2008-02-11-drought_N.htm

INDEX

D

G

H

I

J

L

M

N

O

S

Y